ELECTRIC VEHICLES AND THE END OF ICE AGE

2ND EDITION

ANUPAM SINGH

© *ELECTRIC VEHICLES*
Anupam Singh
2nd Edition

Publication Date: November 2022
Price: ₹ 595$ 14.99
ISBN: *978-93-95673-62-4*
Published by:
Adhyyan Books
Office No. 125,
Opposite Vivanta by Taj,
DDA SFS. Pocket-1, Dwarka,
Sec-22, New Delhi-110077
Website: http://adhyyanbooks.com
E-mail: contact@adhyyanbooks.com

Printed at: Repro Printers, Delhi

CONTENTS

Dedicated to my dearest children,

Vadaanya & Shivtosh

May your future be brighter,

electric and cleaner

—Papa

PREFACE

The first edition of this book, ***Electric Vehicle and the end of the ICE Age***, was published in April 2019. In the preface, I mentioned, "*The pace of change in the EV space is so fast that by the time this book goes to print, some of the information will already be outdated.*" I'm not a soothsayer, but I do take educated guesses. And I've been proved right.

More than three years have passed since the first publication. And the changes have been fast-paced, more than what I had anticipated. So let's have a look at where the EV space stands today.

Global EV sales volumes crossed 6.5 million units in 2021, a growth of 108% over 2020. That's more than doubling sales in just one year! In 2019, the year of the first publication of this book, the EV sales volumes were a little over 2.2 million. So, in 2 years, the volumes went up three times!

And the party is just starting.

The share of EVs in global light vehicle sales reached 8.3% in 2021, compared to 4.2% in 2020. That means EVs not only doubled the volumes in 2021, but they also doubled their market share.

The picture is even rosier for buses – electric buses formed 44% of total sales in 2021—the electric 2 & 3-wheelers following closely with 42% of total sales.

Two things stand out in this remarkable trajectory :

1. Tesla led the EV growth with 936,000 deliveries. Out of this, Tesla Model 3 sales stood at 501,000 units, continuing its record as the bestselling EV since 2018. It ranked No. 2 among all mid-size car nameplates, ICEs included. Only Toyota's Camry sold more units.
2. The sale of EVs in China jumped by more than 2 million units, more than the volume increase of all other regions combined. China today accounts for more than 50% of global EV sales.

The above figures pertain to passenger, light, and commercial vehicles and do not include electric buses and larger commercial vehicles. But that would not be a significantly large number.

What is critical to see is the split between Battery Electric Vehicles (BEVs) and Plug-in Hybrid vehicles (PHEVs). In 2021, BEVs stood for 71% of total EV sales, while PHEVs formed the rest at 29%. So the customer choice is loud and clear – they want pure electric vehicles.

Also important is to see the variations in different market regions. Which are the markets growing the fastest? From where is the subsequent growth expected to come?

EV sales in China had a hyper-growth of 155% in 2021. In North America, the EV volumes surged 96%, albeit on a low base of 2020. The European region witnessed a growth of 66% in 2021 for EVs, and these are the main markets leading the charge for EVs.

In Europe, the EV share in total sales increased from 10% to 17%, with a peak of 26% in December 2021. This expansion is massive – one in four new European vehicles sold was an EV. In China, the EV share increased from 5.5% to 13.3% - an enormous jump.

There was high growth in some other countries, notably South Korea, with sales of over 114,000 EVs. Israel, Australia, India, and Japan clocked sales of over 10,000 units each.

What we are witnessing now is a scorching pace of EV growth led by some key markets worldwide. And although Tesla ranks No. 1 in EV sales for the year 2021, legacy automakers like Volkswagen, General Motors, and Ford are catching up fast. BYD of China has overtaken Tesla in the first six months of 2022.

The writing is on the wall: "Sales of internal combustion vehicles will be in a state of permanent decline."

As per a recent Bloomberg New Energy Finance (BNEF) report, in 2025, the global auto industry will sell 20.6 million plug-in vehicles, of which 75% will be fully battery-powered vehicles. This sales growth will take the share of EVs to 23% of new passenger vehicle sales globally.

China and Europe will account for 80% of EV sales in 2025, with adoption running much slower elsewhere. Most importantly, the market is shifting from being driven primarily by policy to one where organic customer demand is the most crucial factor.

By far, the most extensive penetration of EVs is in Norway. In 2021, electric cars accounted for 72% of new car sales in Norway. Sweden came in at a distant second with EV penetration of 45%, followed by the Netherlands at 30%. But in absolute terms, the largest electric car market in Europe was Germany, where EVs constituted 25% of total sales in 2021.

Electric vehicles have become the road transport technology for many governments and the automotive industry. The US government announced in November 2021 an ambitious 50% electrification target for new cars by 2030, supported by the installation of 500,000 charging points to help increase consumer confidence. In Europe, the EU Commission proposed to bring the CO2 emission standard for new cars to zero by 2035.

At the same time, several automakers announced electrification targets. For example, Volkswagen said half of its sales would be electric by 2030, and Ford said it expects 40%-50% of its sales to be electric by the end of the decade. Another significant milestone in 2021 was Toyota's statement, the largest car manufacturer announcing new investments to achieve electric car sales of 3.5 million a year by 2030.

As new roadmaps are rolling out for the electrification of road transportation, one thing is clear – the auto industry is in the throes of a significant revolution.

Even in 2022, sales of electric vehicles (EVs) are booming. The first-half results are in, and they rock. Despite supply chain issues and higher upfront costs, the auto industry reports strong EV performance. As per reports, 4.3 million EVs were sold worldwide in the first half of 2022, registering a growth of 62% compared to the same period last year.

EV sales pattern in the first half of 2022 is the same as in 2021 – China and Europe dominate the electrification of cars.

2.4 million EVs were delivered to customers in Mainland China in H1 2022. This number equated to 26% of all passenger cars, against just 10% in H1 2021. So China continues to be the world's largest and fastest-growing EV market. 1.1 million EVs were delivered to European customers, accounting for 20% of all passenger cars, and EVs represented 16% in H1 2021. In comparison, the US is still catching up, with 414,000 Evs delivery to customers in H1 2022. But EVs grew from 3% of new cars in H1 2021 to 6% of new vehicles in H1 2022, a rise of 62%.

The era of electric vehicles has arrived. The following 4-5 years will see further acceleration in EV penetration as legacy automakers and new players launch new electric cars, giving a broader choice to customers. As the EVs approach price parity with their ICE counterparts in the not-too-distant future, the transition to electrics will be even smoother.

In the second edition of this book, I have updated all the chapters with the latest information available when sending it for publication. No doubt, in a fast-paced, dynamic world, the data will become outdated pretty soon. However, the direction of change will surely be the same.

This book begins at the beginning - when electric vehicles dominated the automotive industry.

The ensuing chapters give an overview of electric vehicle technology, including batteries and charging equipment. Next, it explores the impact electric vehicles have on the environment, oil demand, and automobile dealerships.

No book on electric vehicles would be complete without a chapter on Tesla because it is Tesla that has turned the automotive industry on its head. I have put together what the legacy automakers plan for their future electrification strategy.

I have taken two specific countries as case studies for EVs--Norway and China. They are pretty different in size, population, customer needs, etc. There are important lessons to be drawn from the policies and experiences of these two countries, which can shape the future of electric vehicles in other countries, including India.

In the end, I have given some guidelines on what to look for while buying electric vehicles and a brief of the world's bestselling electric cars.

I have used the distance measure of kilometers (km) and miles interchangeably throughout the book. The conversion ratio is 1 mile = 1.61 km. Other useful conversion ratios are : 1 gallon (US) = 3.78 litres, 1 horsepower (hp) = 0.735 kilowatt (kW)

After reading the book, I sincerely hope you will become more enlightened about electric vehicles and share your newfound knowledge with your friends, family, and colleagues.

I would love to connect with you and exchange thoughts, ideas, information, snippets, news, and other exciting things on electric vehicles. My coordinates are as follows:

Email: anupamsingh.ev@gmail.com
Twitter: @TygrAnupam
Facebook: fb.me/anupamEV
LinkedIn: linkedin.com/company/electric-vehicles-book
Website: http://www.electricvehiclesage.com
Instagram: instagram.com/anupamev

Anupam Singh
September 2022

List of Abbreviations

Abbreviation	*Full Form*
2DS	2 Degree Scenario
AC	Alternating Current
Ah	Ampere hour
AWD	All-wheel Drive
B2DS	Beyond 2 Degree Scenario
BEV	Battery Electric Vehicle
BLDC	Brushless Direct Current
BMS	Battery Management System
BNEF	Bloomberg New Energy Finance
BTU	British Thermal Unit
CARB	California Air Resources Board
CII	Confederation of Indian Industry
CO2	Carbon Dioxide
CS	Charging Station
CVT	Constantly Variable Transmission
DC	Direct Current
DOD	Depth of Discharge
ECU	Engine Control Unit
EMF	Electromotive Force
EOL	End of Life
ESS	Energy Storage System
EU	European Union
EV	Electric Vehicle
EVSE	Electric Vehicle Supply Equipment
FC	Fuel Cell

FCEV	Fuel Cell Electric Vehicle
GHG	Greenhouse Gas
GM	General Motors
GWh	Gigawatt Hour
HEV	Hybrid Electric Vehicle
hp	Horsepower
ICE	Internal Combustion Engine
IEA	International Energy Agency
IM	Induction Motor
IMF	International Monetary Fund
IRENA	International Renewable Energy Agency
kmph	Kilometers per hour
kW	Kilowatt
kWH	Kilowatt Hour
LCC	Life Cycle Costs
LCV	Light Commercial Vehicle
LDV	Light Duty Vehicle
Li-Ion	Lithium Ion
LFP	Lithium Ion Phosphate
mbd	Million Barrels per Day
MPGe	Miles per Gallon equivalent
MUV	Multi-utility Vehicle
NdFeB	Neodymium Iron Boron
NEV	New Energy Vehicle
Ni-Cd	Nickel-Cadmium
Abbreviation	Full Form
NiMH	Nickel-Metal Hydride
Ni-Zn	Nickel-Zinc
Nm	Newton-metre

OECD	Organization for Economic Co-operation & Development
OEM	Original Equipment Manufacturer
OPEC	Organization of the Petroleum Exporting Countries
PHEV	Plug-in Hybrid Electric Vehicle
PLDV	Passenger Light Duty Vehicle
PM	Permanent Magnet
PWM	Pulse Width Modulation
PV	Photo Voltaics
RE	Renewable Energy
RPM	Rounds per Minute
SLI	Starting, Lighting & Ignition
SOC	State of Charge
SOH	State of Health
SRM	Switch Reluctance Motor
SUV	Sports Utility Vehicle
SynRM	Synchronous Reluctance Motor
TCO	Total Cost of Ownership
TMS	Thermal Management System
TWh	Terawatt hour
UC	Ultra Capacitor
VAT	Value Added Tax
VGI	Vehicle-to-Grid Integration
Wh	Watt Hour
W-IWM	Wireless In-Wheel Motor
WTW	Well-to-Wheel
ZEV	Zero Emission Vehicle

1

HISTORY OF ELECTRIC VEHICLES

"History happens twice because people don't listen the first time."

—Anonymous

The din of electric vehicles (EVs for short) is fast reaching a crescendo. Everyone in the automotive field (and outside of it) is talking about EVs at the drop of a hat. From initial skepticism, the narrative is turning into a sure-footed march towards the inevitable shift of the vehicle power plant away from fossil fuels. The future seems unavoidable--and it is electric.

But what about the past? Where did it all begin?

For those of you who think that the electric vehicle is a modern or recent occurrence--I've got news for you. EVs *preceded* Internal Combustion Engine (ICE) vehicles. Yes! The year 1830 witnessed the development of the first crude EVs, and the first motor vehicle powered by ICE rolled out much later in 1885.

Naturally, one might wonder what would have happened to EVs if they existed long ago. It is helpful to look back and find out the story's origins to get a perspective because history is about to repeat itself.

So let's delve into the past.

Flashback!

Early Days

A tiny model car powered by an early type of electric motor was built in 1828 by Anyos Jedlik of Hungary. It was more of a toy, but it was enough to be a proof of concept. Between 1832 and 1839 (the exact year is uncertain), Robert Anderson of Scotland invented a crude electric-powered carriage. Then in 1834, Thomas Davenport, a blacksmith in Vermont (USA), installed a battery-powered DC Motor in a small model car which he operated on a short circular electrified track. The same year, Professor Sibrandus Stratingh of Groningen, the Netherlands, and his assistant, Christopher Becker, created a small-scale electrical car powered by non-rechargeable primary cells.

A Scottish chemist Robert Davidson built the first known real-life electric car in 1837, and Zinc-acid batteries or galvanic cells powered it. Another Scotsman, Robert Anderson, invented the first crude electrical carriage between 1832 and 1939, powered by non-rechargeable primary cells.

It wasn't until 1859 that a viable means of storing electricity onboard a vehicle came into being when the French physicist Gaston Plantè invented a rechargeable lead-acid battery. Another French scientist, Camille Alphonse Faure, improved the design of the battery significantly in 1881, which led to their manufacture on an industrial scale and great success in early electric cars.

Cut to 1884, when an English inventor, Thomas Parker, built the first production electric car in London. His company, Elwell-Parker Ltd., which later consolidated with other companies to form the Electric Construction Corporation, produced almost every conceivable piece of electric equipment, including dynamos, motors, controllers, alternators, accumulators, transformers, etc. This company manufactured the first electric 'dog cart' in 1896 and had a virtual monopoly on the British electric car market in the 1890s.

France and the United Kingdom were the first nations to support the overall development of electric vehicles. Andreas Flocken, an engineer, built the first electric car in 1888 in Germany.

The first successful electric car in the United States was built in Des Moines, Iowa, by a Scotland-born chemist named William Morrison in 1891. Twenty-four lead-acid storage cells powered this car, which was stored under the seats and needed charging every 50 miles. The 4-horse power vehicle had a top speed of 20 mph and may well have been the first land vehicle steered with a wheel.

The Electrobat was another successful electric automobile introduced in 1894. It was designed and built by mechanical engineer Henry G. Morris and chemist Pedro G. Salom in Philadelphia, Pennsylvania. In 1896, building off the Electrobat, Morris and Salom constructed their two-seater Electric Road Wagon. They formed the Electric Carriage and Wagon Company, evidently the first electric car company in America. To increase the usage of The Electric Road Wagon in New York, the company converted these cars into a series of coupes and hansoms.

These vehicles had rear-wheel steering, two 1/2 hp motors, and 44 lead-acid cells, giving them a range of 30 miles.

The Heydays

Electric vehicles took off in the late 1890s and early 1900s. Remember that this time, in 1885, Karl Benz of Germany designed and built the world's first practical automobile to be powered by an internal combustion engine. So, the competitor to electric-powered vehicles had already been born but was still in its infancy.

In 1897, The London Electric Cab Company began regular service using cars designed by Walter Bersey, which used a 40-cell battery and a 3-hp electric motor with a 50-mile range between charges. In the same year, the first electric taxis hit the streets of New York when Samuel's Electric Carriage and Wagon Company began running 12 electric cabs.

In 1898, the legendary Dr. Ferdinand Porsche, at age 23, built his first car, the Lohner Electric Chaise. With a hub-motor at each driving wheel, it was the world's first front-wheel-drive vehicle. Porsche's second car was a hybrid, using an internal combustion engine to spin a generator that provided power to electric motors located in the wheel hubs. On battery alone, the car could travel nearly 40 miles.

Now you know that Porsche began its journey with an electric car!

Despite the slow speed limited to 32 kmph, electric vehicles had several advantages over their early-1900s competitors. They did not have the vibration, noise, smell, manual start, and gear changes associated with gasoline cars. The gasoline car still has them when the cycle is turning full circle!

Electric cars found popularity among well-heeled customers who used them as city cars, where their limited range proved to be even less of a disadvantage. Electric cars were often marketed as suitable vehicles for women drivers due to their ease of operation. The perception of "women's car" stigmatized these electric cars, leading some companies to affix radiators to the front to disguise the car's propulsion system!

Early electric vehicles, such as the 1902 Wood's Phaeton, were little more than electrified horseless carriages and surreys. The Phaeton had a range of 18 miles, a top speed of 14 mph, and cost $2,000. While basic electric cars cost under $1,000, most early electric vehicles were ornate, massive carriages designed for the upper class. They had fancy interiors with expensive materials and averaged $3,000 by 1910.

The lack of power infrastructure led to the slow growth of these electric cars. But by 1912, houses started getting wired for electricity, creating a sudden demand for electric vehicles.

In the United States, by the turn of the century, 40 percent of automobiles were powered by steam, 38 percent by electricity, and 22 percent by gasoline. A total of 33,842 electric cars got registered in the United States, and America became the country where electric vehicles gained the most acceptance.

Electric automobiles began to hold speed and distance records. On December 18, 1898, Count Gaston de Chasseloup-Laubat of Paris,

France, became the fastest man alive. Mr. count had driven his speed demon electric vehicle Jeantaud to speeds never known or experienced by any human. This speed was phenomenal (by the times' standards), 39 miles per hour. Hair raising! This auspicious event earned Count Gaston the nickname "the Electric Count."

The record would last just a few days before falling into a faster electric vehicle. On April 29, 1899, a Belgian named Camille Jenatzy reached a top speed of 65 miles per hour (105 kmph) in his ' rocket-shaped vehicle, La Jamais Contente, which in English means 'Never Satisfied.'

Also, in 1899, Walter Baker established the Baker Motor Vehicle Company and began manufacturing Brass Era electric automobiles in Cleveland, Ohio. Baker made many contributions to automotive design, including the first shaft drive, the first use of alloy steel, the first "full floating" ball-bearing rear axle, the first horseless delivery wagon, and the first use of aerodynamic streamlining.

The decline

The fall of electric vehicles was brought upon by many developments, each development giving a deadly blow to battery-operated automobiles one after the other in quick succession.

By the 1920s, the improved road infrastructure required vehicles with a more fantastic range than electric cars. The Discovery of enormous petroleum reserves worldwide led to the availability of cheap gasoline. Charles Kettering dealt a lethal blow with the invention of the electric starter in 1912, eliminating the need for the cumbersome hand crank required for internal combustion cars.

The final blow was perhaps given by Henry Ford with the initiation of mass production of internal combustion engine vehicles, making these vehicles much cheaper than electric ones and widely available. The iconic petrol-powered Ford Model T was introduced in the market in 1908, changing the course of automobile history forever. In 1912, an electric roadster sold for $1,750, while a gasoline car sold for $650.

Most electric carmakers stopped production at some point in the 1910s. By the 1920s, the heyday of electric cars had passed. And by 1935, electric vehicles had disappeared.

The years following until the 1960s were dead years for electric vehicle development and used as personal transportation.

The slow resurgence

As the ICE development progressed rapidly, electric vehicle technology stagnated.

In 1959, Henney Coachworks and the makers of Exide batteries introduced an electric car called Henney Kilowatt in 36-volt and 72-volt configurations. The 72-volt models had a top speed of 96 kmph and could travel for an hour on a single charge. But consumers found it too expensive compared to gasoline cars, and the production ended in 1961.

During the 1960s, interest in EVs increased due to environmental awareness and concerns about oil dependency. Many attempts to produce practical EVs occurred from 1960 to the present.

Among the developments was the Battronic electric truck in 1964 and the Sebring-Vanguard CitiCar in 1974, both in the US. The

Battronic truck was capable of speeds of 40 kmph, a range of 100 km, and a payload of 2,500 pounds. The CitiCar was a miniature commuter car with a top speed of 70 kmph and an 80-100 km range. Production of about 2,000 CitiCars happened, and it remained the most-produced American electric car until 2011, when the Tesla Roadster surpassed it.

Another contemporary to the CitiCar was the Elcar, also known as the Zagato Zele. It was a small electric car produced by the Italian company Zagato, and the tiny vehicle could reach a speed of 70 kmph and had a range of 100 km when fully charged.

The American Motors Corporation (AMC) led the EV development through its associations with Sonotone Corporation for nickel-cadmium batteries in 1959 and then with Gulton Industries for a new battery based on lithium in 1967. In 1975, the US Postal Service purchased 350 electric delivery Jeeps from AM General, a division of AMC.

While the EVs were still facing oblivion on planet Earth, an electric car found its place under the sun on July 31, 1971, becoming the first human-crewed vehicle to drive on the Moon! The car was the Lunar Rover, developed by Boeing and GM subsidiary Delco Electronics. Apollo 15 mission saw the first deployment of this electric vehicle.

Revival of Interest

The energy crisis of the 1970s and 1980s forced a renewed interest in electric cars as a way to escape the fluctuations of the hydrocarbon energy market. In 1990, the chairman of GM, Roger Smith, unveiled Impact, a concept electric car, and announced that

GM would build electric cars for sale to the public. Also, in the early 1990s, the California Air Resources Board (CARB) began a push for lower-emission vehicles, with the ultimate goal being a move to zero-emission vehicles.

The tide was beginning to turn slowly but surely.

At the beginning of 1996, GM produced 1,117 units of its EV1, and the car was only available for people in California, Arizona, and Georgia. People could not buy these electric cars as the company did not put them on sale. But people could use them on lease.

The car boasted a range of about 100 miles on a single charge and could go from zero to 60 in just seven seconds. While consumers responded positively to the EV1, it wasn't a profitable business for GM, and the company decided to recall all of the vehicles once leases had expired. The company then destroyed most cars, keeping only 40 models to donate to museums and other institutions.

GM's EV1, Chrysler's TEVan, Ford's Ranger EV, Honda's EV Plus hatchback, Nissan's Altra EV miniwagon, and Toyota's RAV4 EV were produced in limited numbers for their customers in California. However, the manufacturers came in for sharp criticism for failing to promote their EVs while joining oil industry lobbyists opposing CARB's mandate.

The makers of the EVs themselves crushed the first battle of revival. Almost all EV"s production was withdrawn from the market, in some cases even destroyed by the manufacturers.

The movie *Who Killed the Electric Car? was produced in 2005-2006* and released theatrically by Sony Pictures Classics in 2006. The film explores the roles of automobile manufacturers, the oil industry, the U.S. government, batteries, hydrogen vehicles, and consumers. Each of their roles in limiting the deployment and adoption of this technology.

Throughout the 1990s, interest in fuel-efficient or environmentally friendly cars declined among consumers in the United States. Instead, they favored sport utility vehicles, which were affordable to operate thanks to lower gasoline prices despite their poor fuel efficiency. Domestic U.S. automakers focused their product lines around truck-based vehicles, which enjoyed larger profit margins than the smaller cars, generally preferred in places like Europe or Japan.

There was a lack of automaker's participation in the electric car, but still, the industry gave several small companies. They produced mainly low-speed, low-range neighborhood electric vehicles (NEVs). Some of the names of such cars are Global Electric Motorcars (GEM), Buddy, and Th!nk City.

Reva Electric Car Company was a new entrant, established in India in 1994 as a joint venture between the Maini Group of India and AEV of California. It launched an all-electric microcar REVAi, also known as G-wiz i, in the UK in 2001 after seven years of research and development.

In 1997, Toyota introduced Prius, the world's first mass-produced hybrid car, which is famous even today. In 1999, Honda's Insight hybrid car became the first hybrid to be sold in North America since the little-known Woods hybrid of 1917.

However, it wasn't until the launch of Tesla's Roadster in 2008 that electric cars suddenly began to sizzle. The Roadster was the first highway-legal serial production. All-electric vehicles use lithium-ion battery cells and are the first production all-electric car to travel more than 200 miles (320 km) per charge. In the same year, China's BYD Auto released F3DM, the world's first plug-in hybrid.

The Mitsubishi i-MiEV was launched in Japan for fleet customers in July 2009 and for individual customers in April 2010. The i-MiEV was launched in Europe in December 2010 and included a rebadged version sold in Europe as Peugeot iOn and Citroën C-Zero. The market launch in America began in Costa Rica in February 2011, followed by Chile in May 2011.

Not to be left behind, in 2010, Nissan released its all-electric Leaf. Nissan. Leaf became the first modern all-electric, zero tailpipe emission, five-door family hatchback produced for the mass market from a major manufacturer.

Only six years later, in 2016, GM launched its first all-electric car, the Chevy Bolt.

The EV revival was now taking serious proportions, and more and more manufacturers started offering electric car models.

In 2011, Smart electric vehicles like Wheego Whip, LiFe Mia electric, Volvo C30 Electric, and the Ford Focus Electric were made available for retail customers. The BYD e6, released initially for fleet customers in 2010, began retail sales in Shenzhen, China, in October 2011. The Bolloré Bluecar was released in December 2011 and deployed for use in Autolib's car-sharing service in Paris.

Models released to the market in 2012 and 2013 include the BMW ActiveE, Coda, Renault Fluence Z.E., Tesla Model S, Honda Fit EV, Toyota RAV4 EV, Renault Zoe, Roewe E50, Mahindra e2o, Chevrolet Spark EV, Mercedes-Benz SLS AMG Electric Drive, Fiat 500e, Volkswagen e-Up!, BMW i3, and Kandi EV.

The Renault–Nissan Alliance reached global sales of 100,000 all-electric vehicles in July 2013. In May 2015, global sales of highway-legal all-electric passenger cars and light utility vehicles passed the 500,000 unit milestone, accounting for sales since 2008. Out of these, Nissan accounted for about 35%, Tesla Motors for about 15%, and Mitsubishi for about 10%.

By mid-September 2015, the global stock of highway-legal plug-in electric passenger cars and utility vans passed the one-million sales milestone, with pure electrics capturing about 62% of global sales.

In December 2016, Norway became the first country where 5% of all registered passenger cars were plug-in electric cars. In September 2018, the Norwegian market share of all-electric cars reached 45.3% and plug-in hybrids 14.9%. For a combined market share of the plug-in car segment of 60.2% of new car registrations that month, becoming the world's highest-ever monthly market share for the plug-in electric passenger segment in Norway and any country.

The Tesla Model 3's 100,000 car got delivered in October 2018. In the same year, this model got listed as the world's bestselling plug-in electric car. This vehicle ended 2019 as the world's bestselling plug-in electric car for the second consecutive year, with just over 300,000 units delivered. It surpassed the Nissan Leaf in early 2020 to become the world's bestselling electric car, with more than

500,000 total units sold by March 2020. Tesla also became the first auto manufacturer to produce 1 million electric cars in March 2020. Global sales of the Model 3 passed the 1 million business in June 2021, becoming the first electric car model to achieve the same.

The number of EVs on the road has exploded. From negligible in 2010 toapproximately 1 million in 2016. And by the end of 2020, there were as many as 10 million electric cars on roads across the globe.

Despite the global downturn in vehicle sales due to the pandemic, EV sales doubled in 2021 to set a new record of 6.6 million.

On this note, our journey in the history of EVs ends. Let us reboard the time machine and be ready to tour the future!

History of Electric Vehicles – A timeline

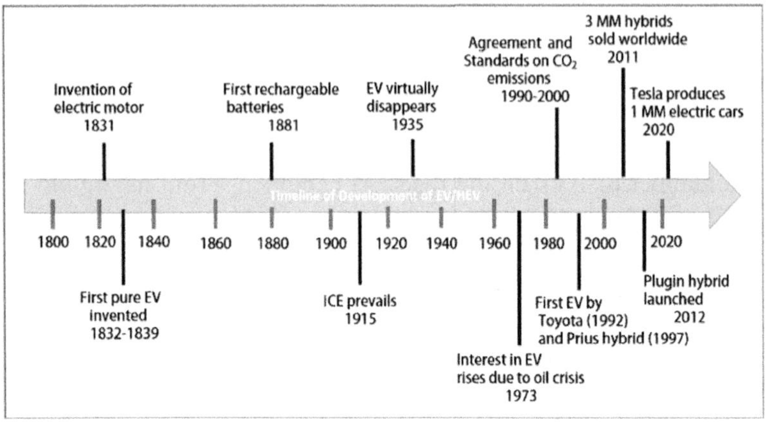

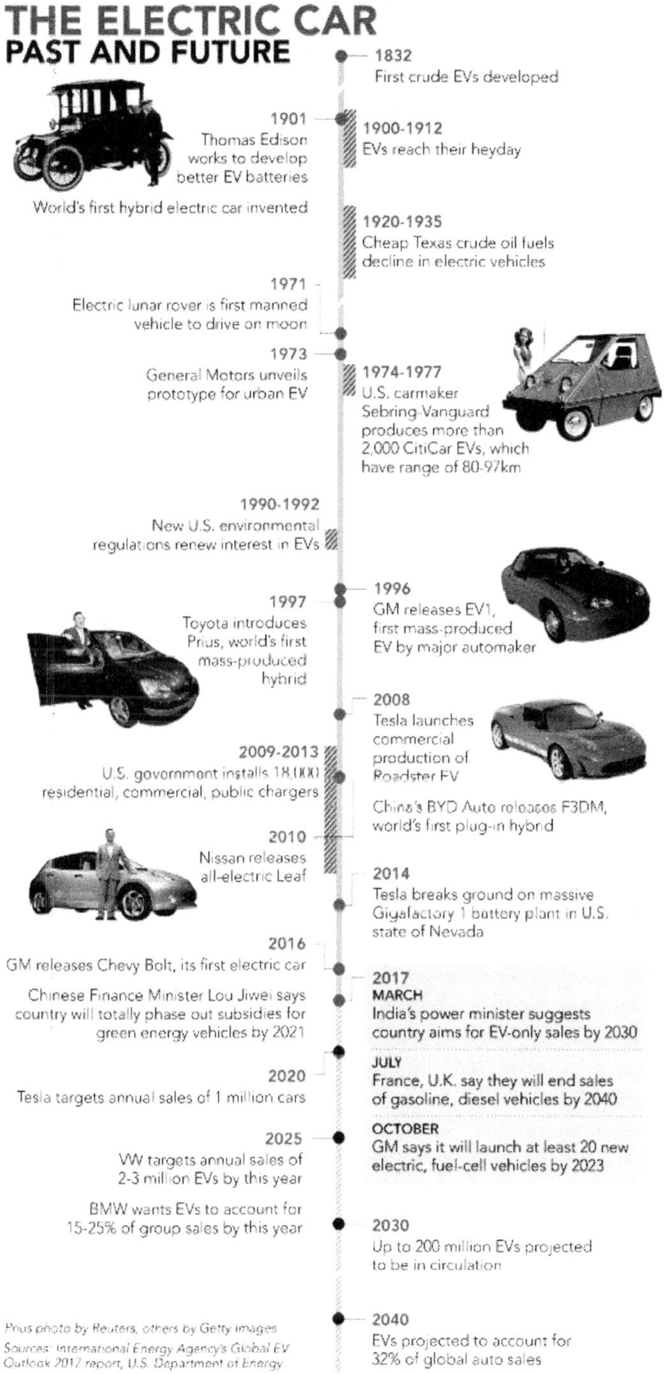
THE ELECTRIC CAR
PAST AND FUTURE
1832
First crude EVs developed
1900-1912
EVs reach their heyday
1901
Thomas Edison works to develop better EV batteries
World's first hybrid electric car invented
1920-1935
Cheap Texas crude oil fuels decline in electric vehicles
1971
Electric lunar rover is first manned vehicle to drive on moon
1973
General Motors unveils prototype for urban EV
1974-1977
U.S. carmaker Sebring-Vanguard produces more than 2,000 CitiCar EVs, which have range of 80-97km
1990-1992
New U.S. environmental regulations renew interest in EVs
1996
GM releases EV1, first mass-produced EV by major automaker
1997
Toyota introduces Prius, world's first mass-produced hybrid
2008
Tesla launches commercial production of Roadster EV
China's BYD Auto releases F3DM, world's first plug-in hybrid
2009-2013
U.S. government installs 18,000 residential, commercial, public chargers
2010
Nissan releases all-electric Leaf
2014
Tesla breaks ground on massive Gigafactory 1 battery plant in U.S. state of Nevada
2016
GM releases Chevy Bolt, its first electric car
Chinese Finance Minister Lou Jiwei says country will totally phase out subsidies for green energy vehicles by 2021
2017
MARCH
India's power minister suggests country aims for EV-only sales by 2030
JULY
France, U.K. say they will end sales of gasoline, diesel vehicles by 2040
OCTOBER
GM says it will launch at least 20 new electric, fuel-cell vehicles by 2023
2020
Tesla targets annual sales of 1 million cars
2025
VW targets annual sales of 2-3 million EVs by this year
BMW wants EVs to account for 15-25% of group sales by this year
2030
Up to 200 million EVs projected to be in circulation
2040
EVs projected to account for 32% of global auto sales
Prius photo by Reuters, others by Getty images
Sources: International Energy Agency's Global EV Outlook 2017 report, U.S. Department of Energy

2

ELECTRIC VEHICLE TECHNOLOGY

"Any sufficiently advanced technology is indistinguishable from magic."
—Arthur C. Clarke

From the outside, you would probably have no idea that a car is electric. When you drive an electric vehicle, the only thing that clues you to its true nature is that it is nearly silent.

But under the hood, there are a lot of differences between ICE and electric cars :

- The electric motor replaces the internal combustion engine
- The electric motor gets its power from the controller
- The controller gets its power from an array of rechargeable batteries

The electric motor, the controller, and the batteries--are the heart of the electric vehicle. The transmission system and the wheels are the other components of the vehicle propulsion architecture, just like in conventional cars. The brakes, steering, air conditioning, energy management, and other auxiliary systems make up the rest of the engineering. All these are, of course, encased in a body structure supported by a frame and suspension.

How does it work?

In simple-to-understand terms, here's how an electric vehicle works :

1. The battery pack stores electrical energy when you charge the battery, and this stored energy gives power to the electric motor and other accessories/components.
2. A controller manages the flow of electrical energy to the motor.
3. The controller modulates the flow of electricity depending on the signals received from the accelerator pedal.
4. The electric motor receives power from the battery and converts it into mechanical energy.
5. The transmission transfers the mechanical energy from the motor to drive the wheels.
6. Some EVs produce regenerative braking energy, and the energy produced during braking or deceleration of the vehicle goes back to the battery pack.
7. The charging port charges the battery pack. The on-board charger converts AC Alternating Current) to DC (Direct Current).

8. You can charge the battery by connecting an external power source to the vehicle's charging port.
9. An EV also has an auxiliary battery to power the vehicle's accessories and other equipment.
10. A DC-to-DC converter charges the auxiliary battery. It converts high-voltage current from the battery pack to low-voltage current.

Now you may wonder how the electric motor, the new engine in the vehicle, converts the electric energy into mechanical energy required to move the car. The process of conversion is as follows:

- When you press the accelerator pedal, the Power Electronics Controller (PEC) controls the flow of electrical energy to the motor.
- The amount of current required to run the motor varies depending on the input on the accelerator pedal.
- Once the motor receives power, the rotors start spinning and produce mechanical energy.
- The mechanical energy produced by the motor starts rotating the gears or transmission.
- The transmission transfers the mechanical energy to the wheels, which propels the vehicle forward.

How's it different?

There are specific inherent characteristics of EVs, which set them apart from ICE vehicles, and they are :

- Instant and smooth acceleration

- Quiet (inside and outside)
- Minimal maintenance
- Efficient energy usage (electric drives have an efficiency of up to 95% compared to the 35-40% efficiency of internal combustion engines)
- Zero tail-pipe emissions (for full-electric vehicles)
- Regenerative braking capability (explained later in this chapter)

All-electric vehicle components

The all-electric or BEVs are increasing in popularity and today take up the lion's share of the total EV market. The below diagram shows the main components of an all-electric car :

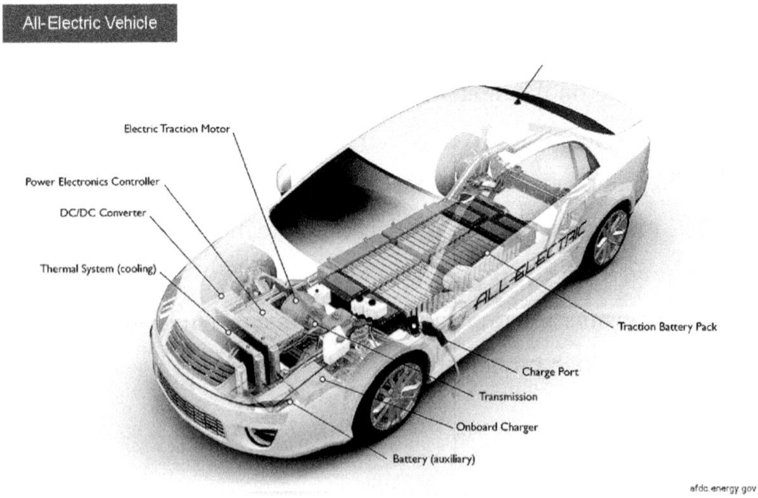

Source: afdc.energy.gov

Let's understand briefly what each component does.

1. Battery pack: The battery stores electrical energy and is the equivalent of a fuel tank in an internal combustion engine. It powers the electric motors of an electric vehicle, and it stores energy in the form of DC. How much energy an EV's battery can hold will depend on its capacity, measured in kilowatt-hours (kWh). The higher the number, the higher the ability, and the farther you can drive that EV on a single charge. The lifetime of a traction battery pack is estimated to be 200,000 miles.

2. Battery Management System (BMS): The Battery Management System (BMS) manages the battery's many cells to operate as if they are a single entity. The EVs battery consists of as little as tens to as many as thousands of mini-cells. Each cell needs to be in a similar condition to the others to optimize the battery's durability and performance. The BMS is often built into the battery's body, though sometimes it is incorporated into the Electric Power Control Unit (EPCU). The BMS mainly oversees the cell's charge/discharge status. Still, when it sees a malfunctioning cell, it automatically adjusts the power status of the cell (on/off) through a relay mechanism (the conditional mechanism for opening/closing other circuits).

3. DC-DC Converter: The traction battery pack delivers a constant voltage. But different components of the vehicle have additional requirements. The DC-DC converter distributes the output power from the battery to a required level. It converts the high-voltage electricity from the EV high-voltage battery into low-voltage (12V) and supplies it to the vehicle's various electronic systems. The high voltage in the battery is converted into low voltage first because all electronic systems in the EV use electricity in low voltage. It also provides the voltage required to charge the auxiliary battery.

4. Electric Motor: Electric traction motor is the main component of an electric vehicle. The motor converts the electrical energy into kinetic energy that moves the wheels. An electric motor is the main component that differentiates an electric car from a conventional vehicle. An essential feature of an electric motor is the regenerative braking mechanism. This mechanism slows down the vehicle by converting its kinetic energy into electric energy and storing it in the battery pack for future use. So, in part, the motor acts as an electric generator. There are two types of motors: DC and AC motors.

An EV's electric motor doesn't have to pressurize and ignite gasoline to move the car's wheels. Instead, it uses electromagnets inside the engine powered by the battery to generate rotational force.

Inside the motor, there are two sets of magnets. One group is attached to the shaft that spins the car's wheels, and the other is inside the housing surrounding that shaft—both the magnets need equal charging so that their polarity is the same and repels one another. The force of the magnets pushing away from one another turns the shaft, spins the wheels, and moves the car forward.

For constant repulsion between the magnets, their polarity has to change as the shaft turns constantly. Otherwise, they would eventually rotate back to where they would attract instead of repelling one another and lock themselves in place. AC power does this automatically, constantly alternating between positive and negative. But since the power from an EV's battery is DC, a device called an inverter is needed to keep flipping the polarity of the magnets.

5. Power Inverter: It converts DC power from the batteries to AC power to control motor speed. The device is responsible for executing acceleration and deceleration, as the processes are crucial in maximizing Ev's driveability. It converts the AC power generated during regenerative braking into DC and is again used to charge the batteries.

An EV's inverter flips polarity quickly, around 60 times per second, to keep the rotational force going. A separate DC converter uses direct power to other vehicle systems (heating, infotainment, and lighting) that don't require alternating currents. The driver can change the frequency of the current sent to the motor, and the higher the frequency, the more frequently the polarity flips. This process generates more rotational force, or torque, and spins the wheels faster.

6. Charge Port: The charge port connects the electric vehicle to an external supply. It charges the battery pack. The charge port is located either in the front or rear part of the vehicle

7. On-board Charger: Onboard charger converts the AC supply received from the charge port to the DC supply for charging the batteries. It also communicates with the charging equipment and monitors battery characteristics such as voltage, current, temperature, and state of charge while charging the pack. The On-board Charger is not needed in fast charging since fast chargers already supply the electricity in direct current.

8. Controller: Power electronics controller determines the working of an electric car. It regulates electrical energy from the batteries to the electric motors. The pedal set by the driver selects the car's speed and frequency of variation of voltage input to the engine. It also controls the torque produced.

9. Auxiliary batteries: Auxiliary batteries are the source of electrical energy for the accessories in electric vehicles. The auxiliary batteries will continue to charge the car without the main battery, preventing the voltage drop produced during engine start from affecting the electrical system.

10. Thermal system (cooling): The thermal management system is responsible for maintaining an operating temperature for the main components of an electric vehicle, such as an electric motor, controller, etc. It functions during charging as well to obtain maximum performance. It combines thermoelectric cooling, forced air-cooling, and liquid cooling.

11. Transmission: This process transfers mechanical power from the electric motor to the wheels through a gearbox. The advantage of electric cars is that they do not require multi-speed transmissions. The transmission efficiency should be high to avoid power loss.

Working on an Electric Motor

Sometime in the not-too-distant future, petrol-powered internal-combustion engines (ICE) will be something you're more likely to find in a museum than an actual car. With the complete takeover of EVs, the motors will replace the engines. It's an excellent idea to demystify the world of electric motors and make yourself familiar with the new technology. Welcome to the world of electromagnetism!

We have primarily developed a reasonable understanding of how combustion powertrains work. Most of us can visualize fuel and air entering a combustion chamber, exploding, pushing a piston down, and rotating a crankshaft that ultimately turns the wheels. Mechanical engineering concepts such as these are comparatively easy to comprehend. But only a minority of us can explain on a bar napkin exactly how invisible electrons turn a car's wheels or how a permanent-magnet motor differs from an AC induction one.

Let me explain the working of an electric motor in simple terms.

In general, motors for EVs work by converting electricity into mechanical energy through the creation of a magnetic field at the fixed part of the machine (the "stator," which is static), whose displacement sets a rotating part (the "rotor") in motion.

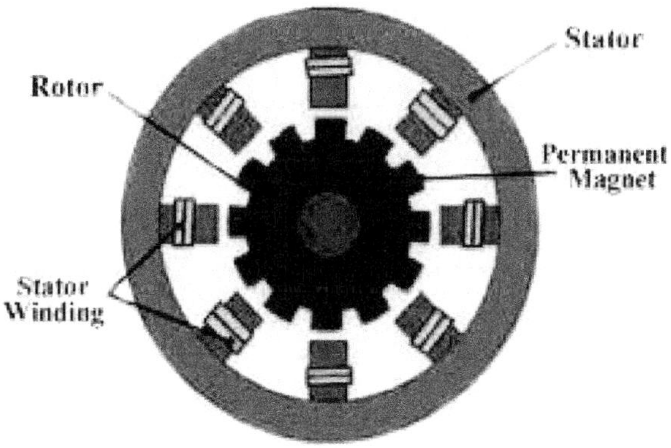

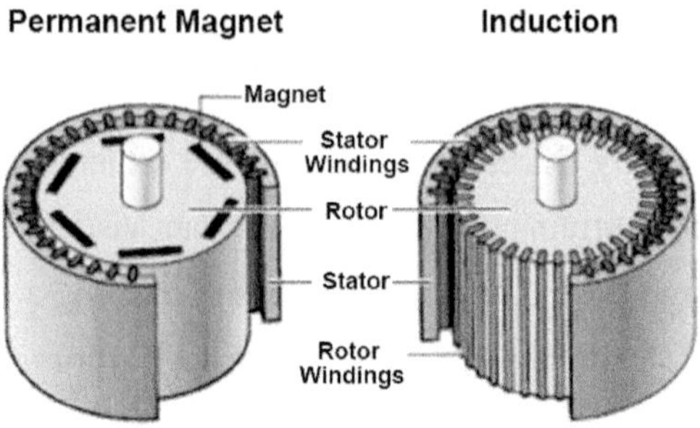

It has to do with magnetism and the natural interplay between electric fields and magnetic fields. When an electrical circuit closes, allowing electrons to move along a wire, those moving electrons generate an electromagnetic field, complete with a north and a south pole. When this happens in the presence of another magnetic field, the opposite poles attract and repel like others.

Electric car motors work by mounting one set of magnets or electromagnets to a shaft and another set to a housing surrounding that shaft. By periodically reversing the polarity (swapping the north and south poles) of one set of electromagnets, the EV motor leverages these attracting and repelling forces to rotate the shaft, thereby converting electricity into torque and ultimately turning the wheels. Conversely—as in the case of regenerative braking—these magnetic/electromagnetic forces can transform motion back into electricity.

Electric motors require alternating currents to spin. Without it, the electromagnetic force would lock their north and south poles together. The continually switching of the north and south cycle keeps an electric car motor spinning.

Nowadays, electric cars can manage AC and DC energy on board, as car manufacturers design their vehicles in such a manner. The battery stores and dispenses DC energy, but the motor needs AC energy again. When recharging the battery, the energy comes into the on-board charger as AC during Level 1 and Level 2 charging and as DC high-voltage current on Level 3 "fast chargers." Sophisticated power electronics handle the multiple on-board AC/DC conversions while stepping the voltage up and down from 100 to 800 volts of charging power to battery/motor system voltages of 350-800 volts to the many vehicle lighting, infotainment, and chassis functions that require 12-48-volt DC electricity.

Types of Electric Motors

Several internal combustion engines are naturally aspirated, turbocharged, direct injection, indirect injection, petrol, diesel, CNG, LPG, etc.

Vehicle applications use different types of electric motors. Based on their design, below are the variety of electric motors used in e-mobility :

- DC Motor (Brushed)
- Brushless DC Motor (BLDC)
- Permanent Magnet Synchronous Motor (PMSM)
- AC Induction Motor
- Reluctance Motor (Switched and Synchronous)

I will not get into details of all types, but just brief you about the two most popular ones: Permanent Magnet Synchronous Motor and AC Induction Motor

Permanent Magnet Synchronous Motor (PMSM)

In Permanent Magnet Synchronous Motors, permanent magnets create a native rotating magnetic field (hence the engine's name, motor) in the rotor.

The rotating magnetic fields of the rotor and stator in these motors are locked, and there is no slip.

The PMSM functions on continuous sinusoidal AC power. This function means that it suffers no torque ripple and needs only one Hall-effect sensor (a type of magnetic sensor) to determine rotor speed and position, so it's more efficient and quieter.

The word "synchronous" indicates the rotor spins at the same speed as the magnetic field in the windings. Its significant advantages are its power density and strong starting torque. The main disadvantage of any EV motor with spinning permanent magnets is that it creates "back electromotive force" (EMF) when not powered at speed, which causes drag and heat that can demagnetize the motor. This motor type also sees some duty in power steering and brake systems, but it has become the motor design of choice in most of today's battery electric and hybrid vehicles.

Hyundai Ioniq 5, Kia EV6, Tesla Model S, 3, X, and Y, uses a permanent magnet motor on the rear axles. VW Group MEB also use them on the rear axles, Jaguar i-pace, Audi e-Tron GT, and Porsche Taycan, just to name a few.

AC Induction Motor

For this motor, we toss out the permanent magnets on the rotor (and their increasingly scarce rare earth materials) and keep the AC power flowing through stator windings as in the PMSM motor.

Standing in for the magnets is a concept that Nikola Tesla patented in 1888, and it was a forerunner of the 3-phase induction motor.

This type of motor works as follows: AC voltage is attached to the copper coils (or windings), and as a result, we get a rotating magnetic field; that field induces a voltage in the rotor, which in turn causes current to flow. That current flow produces its rotating magnetic field in the rotor, which lags behind the stator's magnetic field. The force between the two magnetic fields that end up moving the rotor is called the Lorentz Force. Through sound reduction, the movement of the rotor gets transferred to the car's wheels.

The rotating magnetic field of the rotor and stator is not synchronized; therefore, this motor is called asynchronous. The induction part comes from the rotating magnetic field as the stator induces voltage and current. When we press the accelerator pedal, the rotor’s magnetic field lags slightly behind the stators. When we decelerate and the motor works as a generator (regenerative braking), the rotating magnetic field of the rotor is ahead of the stator. This difference in the rotating magnetic fields is called “slip” and is generally up to 5%, depending on the motor design.

The typical efficiency of a 3-phase asynchronous induction motor used in the automotive industry is around 90%. Because of its robustness, simplicity, longevity, and zero exotic materials requirements, this motor is used almost exclusively in industrial processes. Also, its good overload characteristics make it a perfect on-demand engine. So, AWD electric cars use them as front motors.

Some cars using the asynchronous induction motors are Audi e-Tron SUV, Mercedes-Benz EQC, Tesla Model S, 3, X, and Y on front axles, and VW Group MEB cars use them on front axles as well.

That's it. Now armed with new-found knowledge, you can make an educated choice for your EV. And you can impress your friends and relatives with some limited jargon!

You probably won't have a choice in selecting the type of motor when buying your electric car. But when it comes to types of electric vehicles (EVs), the options people can buy go beyond those powered by batteries alone.

There are different types of EVs available in the market today, and it'll help to familiarize yourself with the broad types. Here's a quick low-down.

EV Types

EVs can either run solely on electric propulsion or have an ICE engine working alongside it – in that case, it is aptly called a 'hybrid.' Combining different drive types reduces the vehicle's overall efficiency and fuel consumption.

EVs can be categorized as follows:

1. Battery Electric Vehicle (BEV)
2. Hybrid Electric Vehicle (HEV)
3. Plug-in Hybrid Electric Vehicle (PHEV)
4. Fuel-cell Electric Vehicle (FCEV)

Let us look at each one of these detail.

Battery Electric Vehicle

BEVs have to rely solely on the energy stored in their battery packs; therefore, the range of such vehicles depends directly on the battery capacity. Anyone familiar with a golf cart (used at airports, too) has experienced a BEV. Typically, they can cover 100-250 km on one charge; however, the top-tier models can go a lot further, from 300 to 500 km.

Once depleted, charging the battery pack takes longer than refueling a conventional ICE vehicle, and it can take as long as 36 hours to replenish the batteries completely. Of course, the charging time depends on the battery capacity. Typically, it takes about 8-11 hours for a full charge when you plug your car into the 230V AC supply of your home or office. There are far less time-consuming options, but none is comparable to the little time required to refill a fuel tank. Done through special DC chargers, fast charging can juice up the batteries in 60-90 minutes.

Hybrid Electric Vehicle

HEVs employ both an ICE and an electric powertrain to power the vehicle. An HEV uses an electric propulsion system when the power demand is low. It is an advantage in low-speed conditions like urban areas; it also reduces fuel consumption as the engine stays off during idling periods, such as traffic jams. This feature also reduces GHG (Greenhouse Gas) emissions.

When a higher speed is needed, the HEV switches to the ICE. The two drivetrains can also work together to improve performance. The manufacturers of electric vehicles use Hybrid power systems to remove turbo lag in turbocharged cars. It also enhances performance by filling the gaps between gearshifts and providing speed boosts when required. The ICE can charge the batteries, and HEVs can retrieve energy utilizing regenerative braking.

Therefore, HEVs are primarily ICE-driven cars that use an electric drivetrain to improve mileage or enhance performance.

Plug-in Hybrid Electric Vehicle

The concept of the PHEV came about to extend the all-electric range of the HEVs. It uses an ICE and an electrical powertrain like an HEV, but the difference between them is that PHEV uses electric propulsion as the main driving force. So, these vehicles require a bigger battery capacity than HEVs.

PHEVs start in 'all electric' mode, run on electricity, and when the batteries are low in charge, it calls the ICE to provide a boost or charge up the battery pack. In this case; the range extends by using ICE. PHEVs can charge their batteries directly from the grid (which HEVs cannot). They have the facility to utilize regenerative braking. PHEVs' ability to run solely on electricity most of the time makes their carbon footprint minor than the HEVs. They consume less fuel and thus reduce the associated cost.

Fuel-cell Electric Vehicle

FCEVs are also known as FCVs (Fuel Cell Vehicles).

They got this name because the heart of such vehicles is fuel cells, and these cells use chemical reactions to produce electricity. Hydrogen is the fuel of choice for FCVs to carry on this reaction. Therefore, they are also known as 'hydrogen fuel cell vehicles.'

FCVs carry the hydrogen in special high-pressure tanks. Another ingredient for the power-generating process is oxygen, which it acquires from the air sucked in from the environment. Electricity generated from the fuel cells goes to an electric motor, which drives the wheels. Storages like batteries or supercapacitors store excess energy in them.

FCVs only produce water as a by-product of the power-generating process, and Tailpipes help in water ejection from the car. An advantage of such vehicles is that they can create their electricity, emitting no carbon and enabling it to reduce their carbon footprint further than any other EV. Another significant advantage of these, and the most important, is that refilling these vehicles takes the same time required to fill a conventional car at a gas pump.

An obstacle to adopting this technology is the scarcity of hydrogen fuel stations, and BEV or PHEV charging stations were not a common scenario even a few years back. Another disadvantage is the high cost of fuel cells, in that they charge more than $200 per kW, which is far greater than ICE (less than $50 per kW).

There are also safety concerns if flammable hydrogen leaks out of the tanks.

FCEVs can outshine BEVs in the future if these obstacles get removed in time. So, don't write them off yet.

Before I end this chapter, I will briefly discuss the EV brake system. It has a massive advantage over the brake system in ICE vehicles simply because of the change in the 'prime mover' of the car.

Brake System

You're wasting energy whenever you step on your car's brakes. As per physics, nothing can destroy power. So, when your car slows down, the kinetic energy propelling it forward must go somewhere, and most of it dissipates as heat and becomes useless. But that energy was still helpful for some other work.

Is there anything to stop energy waste?

Yes, there is! But only with hybrid or fully electric vehicles. The braking system can end this energy wastage. This system recaptures much of the car's kinetic energy and converts it into electricity so that it can recharge the car's batteries. This system is called regenerative braking.

With regenerative brakes, the vehicle's system does most of the braking. When the driver steps on the brake pedal of an electric or hybrid vehicle, these brakes put the vehicle's electric motor into reverse mode, causing it to run backward, thus slowing the car's wheels. While running backward, the engine also acts as an electric generator, producing electricity feeding into the vehicle's batteries. These brakes work better at certain speeds than others and are most effective in stop-and-go driving situations. However, hybrids and fully electric cars also have friction brakes as a backup system when regenerative braking doesn't supply enough stopping power.

Let me explain this in another way.

An electric vehicle has two independent brake systems: the traditional mechanical/ hydraulic brake system and the electric drive motor forms the second as an "engine brake." The advantage of this "engine brake" compared with the combustion engine is that the energy released by the electric motor/generator during braking and deceleration is recovered and fed into the high-voltage battery. This regenerative braking contributes to the high efficiency of electric vehicles, particularly in city traffic. In addition, the wear of the vehicle brakes is reduced by the regenerative braking system.

So, you're in safe hands and in a pretty efficient way. Remarkable, isn't it?

3

BATTERY TECHNOLOGY

"Man's brain is equivalent to an electric battery…a group of electric batteries will provide more energy than a single battery."
—Napoleon Hill

The heart of any vehicle is the power source–in the case of fossil-fuel vehicles, it's the internal combustion engine. That is where the power is generated and then transmitted to the wheels via the driveline. It looks like child's play, but it's unbelievably complicated.

For an electric vehicle, the power source is a rechargeable battery. We are familiar with batteries–in our laptops, cell phones, and all consumer electronic devices. But the batteries used in vehicles are quite different because they are required to handle high power and high-energy capacity within a limited space and weight. High specific energy is needed from a source to provide a long driving range, whereas high specific power helps to increase the acceleration.

The higher the energy density, the more energy can be stored and released. The unit of energy density is watt-hours per kilogram [Wh/kg], and the calculation includes the electrical work [Wh] and the weight [kg] of the battery. The energy density determines the range of an electric vehicle.

Other characteristics of a perfect energy source are fast charging, extended service and cycle life, and less cost and maintenance.

Battery capacity

The high value of specific gasoline energy gives a conventional ICE-powered vehicle a range of 300–400 miles with a full gasoline tank. Gasoline has a theoretical power of 13,000 Wh/kg, which is over 100 times higher than the specific energy of 120 Wh/kg of typical Li-ion batteries. It would be too big and heavy to have a battery pack with the same power as a full tank (e.g., 16 gallons) of gasoline.

However, since electric propulsion is much more efficient than an ICE, less energy is needed to propel an EV. Considering the efficiency of 80% for EV propulsion and 20% for ICE, the total energy stored for an EV can be a quarter of what a regular ICE-powered vehicle needs for the same mileage. Based on the current battery technology, it is not practical to consider a pure BEV with a mile range of 300–400 miles since it would require a battery pack bigger than 100 kWh that can weigh over 900 kg. Nevertheless, it is realistic to have a battery pack of around 30 kWh to achieve a 100-mile range, even based on current battery technologies.

The unit to measure an electric vehicle's battery capacity is kilowatt-hours (kWh), the same unit your home electric meter records to determine your monthly electric bill. In the EV world, kilowatt-hours are to batteries as liters (or gallons) are to fuel tanks. However, a full battery can't be equated entirely with a full fuel tank. It's important to understand that the battery's rated capacity is something you will never be able to use. A "state-of-charge" management system never lets the battery become 100 percent full or 100 percent empty to preserve battery efficiency and life. A more relevant measure might be a battery's usable capacity, but that's enveloped in mystery too.

The manufacturer never reports the unusable capacity because the difference is significant. State-of-charge battery management, an essential feature of modern electric cars, leaves the manufacturer with about 60-70 percent of the rated capacity to work.

In other words, don't rely too much on whether the Nissan Leaf has a lithium-ion battery rated at 24 kWh or the Tesla Roadster's battery is at 54 kWh. You'll never be able to tap all of that energy, anyway. You focus on the specific figure describing a given battery capacity. The consumers should use the rated capacity figure to compare the relative size of batteries.

Charging

Charging comes down to two familiar resources: time and money. The time taken for charging depends on the battery size and the consumer's charger's voltage, and the location and time determine the EV's cost of charging.

Until recharging stations are reliably available in shopping areas and workplaces in sufficient numbers (a development that might take years), most owners will charge their vehicles at home. And most of them will use the 120-volt 'trickle-charging' cord with most EVs. A trickle-charging cable tops off batteries from any standard electrical outlet that is reachable without an extension cord (an extension cord is a big no-no in this process). Unfortunately, it can take more than 20 hours to complete a full charge on 120 volts. Instead, an EV buyer will likely want to install a 240-volt home charger, enabling much faster fills.

The cost of charging, meanwhile, depends on the location where the EV is parked and whether the car is refueling during peak or non-peak hours.

Range

The range is the stickiest question faced by the new EV drivers, and that's because it varies much more for an EV than a conventional car. High and low temperatures affect battery performance and reduce range, and this variation led to the introduction of EVs in areas that aren't very hot or cold. Also, quick acceleration and fast driving discharge the battery faster. Even aggressive braking hurts because it cheats the EV's regenerative braking system of the chance to recapture some energy and recharge the battery.

Because of all these variables, if the manufacturer says 100 miles of range, it could be 60 or 130 miles. You can't tempt fate with the low fuel light as you can in a gasoline car because your garage may be the only fill-up place. So, if you're wrong, you're walking. And that's why there's so much talk about range anxiety.

Basic Terms of Battery Performance and Characterization

Cell, Module, and Pack. A single cell is a full battery with two current leads, a separate compartment holding electrodes, a separator, and an electrolyte. A module comprises a few cells either by physical attachment or welding in between cells. A pack of batteries is composed of modules and placed in a single container for thermal management. An EV may have more than one pack of batteries situated in a different location in the car.

Ampere-hour Capacity. The total charge discharged from a fully-charged battery under specified conditions is known as Ampere-hour (Ah). The Rated Ah capacity is the nominal capacity of a fully- charged new battery under the conditions predefined by the manufacturer. A minor condition, for example, can be defined as 20°C and discharging at 1/20 C-rate. People also use Wh (or kWh) capacity to represent a battery capacity.

The rated Wh capacity is:

Rated Wh Capacity = Rated Ah Capacity x Rated Battery Voltage

C-rate. The nominal C-rate represents a charge or discharge rate equal to the capacity of a battery in one hour. A 1C rate means the discharge current will discharge the entire battery in 1 hour. For a 100 Ah battery, 1C is equal to charging or discharging the battery at 100 Amps. Correspondingly, 0.1C is equivalent to 10 Amps, and 2C for charging or discharging the battery at 200 Amps.

Specific Energy. Specific energy, also called gravimetric energy density, defines how much energy a battery can store per unit of mass. Watt-hours per kilogram (Wh/kg) depicts its unit of measurement.

Specific Energy = Rated Wh Capacity / Battery Mass in kg

The specific energy of a battery is the crucial parameter for determining the total battery weight for a given mile range of an EV.

Specific Power. Specific power, also called the gravimetric power density of a battery, is the peak power per unit mass, and W/kg is its unit of measurement.

Specific Power = Rated Peak Power / Battery Mass in kg

Energy Density. Energy density, also called volumetric energy density, is the nominal battery energy per unit volume (Wh/l).

Power Density. Power density is the peak power per unit battery volume (W/l).

Cut-off Voltage. Cut-off voltage is the minimum allowable voltage defined by the manufacturer. Sometimes it is interpreted as the "empty" state of the battery.

State of Charge (SOC). SOC is the remaining capacity of a battery, and it is affected by its operating conditions, such as load current and temperature.

SOC = Remaining Capacity /Rated Capacity

SOC is a critical condition or a parameter for battery management. Accurate gauging of SOC is very challenging but is the key to the healthy and safe operation of batteries.

Depth of Discharge (DOD). DOD indicates the percentage of the total capacity of discharged battery. Deep-cycle batteries release 80% or higher of DOD.

DOD = 1 - SOC

State of Health (SOH). SOH can be defined as the ratio of the maximum charge capacity of an aged battery to the total charge capacity when the battery is new. SOH is an essential parameter for indicating the degree of performance degradation of a battery and for estimating the battery's remaining life.

SOH = Aged Energy Capacity /Rated Energy Capacity

Cycle Life (number of cycles). Cycle life is the number of charge/discharge cycles the battery can handle at a specific DOD (usually 80%) before it fails to meet specific performance criteria. The actual operating life of the battery is affected by the charging and discharging rates, DOD, and other conditions such as temperature.

The higher the DOD, the shorter the cycle life. A larger battery can be used for a lower DOD during normal operations to achieve a higher cycle life.

Calendar Life. Calendar life is the expected life span of the battery under storage or periodic cycling conditions. It can be strongly related to the temperature and SOC during storage.

Battery Management System (BMS). BMS combines sensors, controllers, communication, and computation hardware with software algorithms designed to decide the maximum charge/discharge current and duration from the estimation of SOC and SOH of the battery pack.

Thermal Management System (TMS). TMS protects the battery pack from overheating, and it also extends its calendar life.

NiMH battery uses Simple forced-air cooling TMS. In contrast, the Li-ion batteries in EVs use more sophisticated and powerful liquid cooling applications.

Battery Life

Like conventional vehicles' engines, the advanced EV batteries are designed for extended life but will wear out eventually. Currently, most manufacturers are offering 8-year/100,000-mile warranties for their batteries.

PlugIn America surveyed Tesla Model S owners, covering 519 vehicles that traveled 13,992,158 total miles – or an average of 17,523 miles per year (more than the 13,476 average annual mileage calculated by the Federal Highway Administration).

It turns out that the Model S lost just about 5% of its power in the first 50,000 miles, at which point battery degradation slowed. Some with 100,000-plus miles had battery degradation of less than 8%. Elon Musk claims to have a Tesla battery pack with over 500,000 simulated miles still running at around 80 percent of its original capacity.

A battery's capacity will degrade as the number of cycles increases. The number of cycles completed measures the battery life. And the industry standard of the processes, close to 80%, is considered a benchmark.

So, what shortens the life of a lithium battery? Several factors:

1. High temperatures
2. Overcharging or high voltage
3. Deep discharges or low voltage

4. High discharges or charge current

The voltage rises slowly when a lithium battery is charging. When it reaches full charge, voltage is at its highest and will not go up anymore. There's a need to keep the voltage from getting excessive power, which is why batteries come with a battery management system (BMS). The system controls the charging voltage, so the maximum charging voltage and the temperature never exceed.

Battery costs

EVs are expensive compared to ICE vehicles, mainly because of the high cost of the battery pack. The battery back constitutes nearly 48% of the total EV price in today's market. At battery prices of $100/kWh, EVs will successfully compete with ICEVs, but mainstream predictions about how and when this will happen to vary widely.

As per Bloomberg New Energy Finance (BNEF), the price of Li-Ion batteries dropped 89% in real terms from $1,200/ kWh in 2010 to $132/kWh in 2021. This steep price decrease is due to technological improvements and economies of scale. Also, fierce competition between major manufacturers has been instrumental in bringing down prices.

In particular, for battery electric vehicle (BEV) packs, prices were $118/kWh on a volume-weighted average basis in 2021. On a regional basis, battery pack prices were cheapest in China, at $111/kWh. Packs in the U.S. and Europe cost 40% and 60% more, respectively.

Prices continued to fall in 2021 as the adoption of the low-cost cathode chemistry known as lithium iron phosphate (LFP) increased, and the use of expensive cobalt in nickel-base cathodes continued to slide. On average, LFP cells were almost 30% cheaper than NMC cells in 2021.

BNEF's *2021 Battery Price Survey* predicts that by 2024 average pack prices should be below \$100/kWh. At around this price point, automakers should be able to produce and sell mass-market EVs at the same price (and with the same margin) as comparable internal combustion vehicles in some markets.

Continued investment in R&D and capacity expansion across the supply chain will help improve battery technology and reduce costs over the next decade. BloombergNEF expects next-generation technologies, such as silicon and lithium metal anodes, solid-state electrolytes, and new cathode material and cell manufacturing processes, to play an essential role in enabling these price reductions.

Types of Batteries

There are many types of batteries used in electric vehicles. The most common variety of battery chemistries are as under :

- Lead-Acid Battery
- Lithium-Ion Battery
- Nickel-Metal Hydride (NiMH) Battery
- Nickel-Zinc (Ni-Zn) Battery
- Nickel-Cadmium (Ni-Cd) Battery

I will discuss only the first two battery categories since EVs mostly use them.

Lead-Acid Battery

The lead–acid battery was invented in 1859 by French physicist Gaston Planté and is the oldest type of rechargeable battery. You are already familiar with it as it is present in every ICE vehicle to provide the high current required by automobile starter motors.

This traditional 12V vehicle electrical system battery has plates made from lead and lead oxide. They both act as electrodes, and sulphuric acid is the electrolyte.

Two primary lead-acid batteries are automobile engine starters and deep-cycle batteries. Automobile alternators provide starter batteries with high charge rates for fast charges. While electric vehicles like forklifts or golf carts use deep cycle batteries as auxiliary house batteries in recreational vehicles, these require different multi-stage charging.

The lead-acid battery has a meager energy-to-weight ratio (30-40 Wh/kg) and a low energy-to-volume ratio. But it is inexpensive compared to newer technologies. Automakers most favored lead-acid batteries for starting, lighting, and ignition (SLI) functions, and the vehicles will not start without them.

Hybrid EVs use lead-acid batteries for SLI functions. One can ultimately have an EV on lead-acid batteries, but the ample space requirement limits its power for a given vehicle size. These batteries are comparatively cheap, widely available, and peppy; golf carts and forklifts mostly use them.

The problem is not only the space and weight; they have fewer cycle lives, and if discharged below 50% of capacity, it shortens the battery life. Then there is the disposal of toxic lead, which harms the environment.

Lead-acid batteries powered such early-modern EVs as the original versions of the EV1 and the RAV4 EV.

Lithium-Ion (Li-Ion) Battery

Lithium-ion (and similar lithium polymer) batteries, widely known for their use in laptops and consumer electronics, dominate the most recent group of EVs in development. Traditional lithium-ion chemistry involves a lithium cobalt oxide cathode and a graphite anode. This chemistry yields cells with an impressive 200+ Wh/kg energy density (twice that of NiMH), strong power density, and 80 to 90% charge/discharge efficiency.

Because of their higher energy density, EVs fitted with Li-Ion batteries can travel longer distances. For example, a Nissan Leaf equipped with a 40 kWh Li-Ion battery pack has a cruising range of 400 km.

The disadvantages of traditional lithium-ion batteries include short cycle lives (hundreds to a few thousand charge cycles) and significant aging degradation. The cathode is also somewhat toxic. Also, traditional lithium-ion batteries can pose a fire safety risk if punctured or charged improperly.

They are expensive, but they have been falling steadily over the last few years and will continue to do so in the coming years.

Lithium-ion batteries comprise a family of battery chemistries that employ various combinations of anode and cathode materials. Each blend has distinct advantages and disadvantages regarding safety, performance, cost, and other parameters. The most prominent technologies for automotive applications are lithium-nickel-cobalt-aluminium (NCA), lithium-nickel-manganese (NMC), lithium-manganese-spinel (LMO), lithium titanate (LTO) and lithium-iron-phosphate (LFP).

The two major battery technologies in EVs are nickel metal hydride (NiMH) and lithium-ion (Li-ion). Because of its mature technology, nearly all HEVs today use NiMH batteries. Due to the potential of obtaining higher specific energy and energy density, adopting Li-ion batteries is expected to multiply in EVs, particularly in PHEVs and BEVs.

The performance of a battery changes as its operating conditions (temperature, charging or discharging current, state of charge (SOC), etc.) and service time vary.

You could think of the electric vehicle battery as the new fuel tank. And while this comparison works on many levels, the analogy doesn't entirely hold. Consumers should understand the three aspects of electric car batteries: capacity, charging, and range, which has been explained earlier in this chapter.

New Battery Technologies

As EV sales are making rapid strides, EV battery technology is advancing to meet market expectations.

Since the early '90s, the battery technology of choice has been lithium-ion, powering everything from phones to EVs to energy grid backup and even satellites. But lithium-ion batteries have some worrying disadvantages.

To begin with, the materials needed to build them, like cobalt, are expensive and environmentally damaging. Cobalt mines use children as their labor, thus targeting the humanitarian ground of society. Then there is a lifecycle problem as these batteries can quickly lose capacity if you repeatedly charge them. Lithium-ion packs are energy-dense but still quite heavy, large, and bulky, limiting electric cars' range. Lithium-ion batteries face fire problems, as a damaged cell can spontaneously catch on fire and burn intensely.

LFP battery – a more robust chemistry

A better alternative to the customarily used Li-ion battery has emerged through years of research and testing. 'LiFEPO4' or 'LFP' battery has safer lithium chemistry, making it better in thermal and structural stability. It also is better in terms of the life cycle – providing up to three times higher lifecycles.

The critical elements of LFP – Iron and Phosphate - are naturally occurring and are available in abundance in India. LFPs have a low self-discharge rate, offering a better range. Additionally, they are less toxic and less harmful to the environment.

LFP is now a proven benign battery type, with advanced chemistry in the first-line generation of batteries available across the globe.

Lithium Sulphur battery – is it the savior?

Here comes the knight in shining armor to save the damsel in distress: lithium sulphur batteries.

On the surface, lithium sulphur seems to solve all lithium-ions problems. It uses far less ecologically harmful materials, can be cheaper to produce, can be up to three times more energy-dense (meaning a lighter battery), and is far less likely to catch fire. All without compromising charge speeds.

So what's the catch?

Well, they have one huge problem. While a lithium-ion battery is useable for around 2000 charge cycles, lithium-sulphur is typically limited to only half. So, a lithium-sulphur battery dies after a year or two of proper use.

So is that it? Is the short story finished?

Hold on – an accidental discovery, which appears now and then out of thin air, could change the world. Just like penicillin, dynamite, X-Ray and microwaves did.

The Drexel team tried new lithium-sulphur approaches to solve this by changing compounds in the battery's cathode.

Their goal was to slow down the chemical reaction that creates polysulphides when the battery charges and discharges. These crystals effectively take sulphur out of the electrode and ultimately cause a massive loss of capacity. Slowing them down could make these very energy-dense batteries last longer.

But they found something incredible: a chemical sulphur phase that stops battery degradation! They were so shocked by the discovery that they had to check 100 times to ensure they weren't misreading it.

This chemical phase, known as monoclinic gamma-phase sulphur, is created in the lab at high temperatures — upwards of 95°C (203°F). This phase, for the first time, saw the room temperature.

This phase completely stops the reaction that creates polysulfides in the battery. This process was so effective that the scientists sent the battery through 4,000 charge cycles without a drop in capacity, meaning it lasted at least twice as long as lithium-ion.

It is also worth noting that their battery was three times as energy-dense as lithium-ion and could charge just as fast!

To say this is a remarkable discovery is a massive understatement. But this new sulphur phase also has other benefits, like reducing battery expansion and increasing safety margins. In other words, this discovery has all the hallmarks of the ultimate mass-market battery. The scientists found it purely by accident.

As with most accidental discoveries, scientists haven't yet figured out what is happening, and they still don't know why this sulphur phase came into existence or how to ensure it stays that way. So further research is needed to answer these questions.

But it will be worth the wait as these batteries will weigh a third of the equivalent lithium-ion batteries and have twice their lifespan!

This discovery means much faster, more efficient EVs with ranges of *thousands of* miles will be commercially viable at a similar cost to today's EVs. What's more, they would still be helpful in the next ten years, dramatically reducing waste and increasing the rate of EV adoption.

In short, lithium-sulphur batteries could allow a vast range of activities to go electric, making net-zero emissions far more feasible.

Promise, promise

Many promising battery chemistries are being explored, with innovations sprouting up quickly. For example, a new silicon-anode lithium-ion battery can charge from empty to full in less than 10 minutes. Designed and built by California-based Enovix, the company claims that the battery maintained 93 percent of its capacity past 1,000 charges and was minimally affected by six months of operation at elevated temperatures.

Enovix isn't the only company investigating the use of silicon as an anode for EV batteries. The USABC (US Advanced Battery Consortium) has already announced funding for similar projects from Amprius, which works on EV batteries with a silicon wire anode. They have also announced funding for Nanoramic, which has developed EV batteries with the silicon-dominant anode, and EV battery equipment can also manufacture them.

Another battery experimenter, Enevate, was developing a porous, sponge-like silicon anode and said it's already working with several EV makers to get its battery, which it claims can go 400km on a five-minute charge, into unspecified 2024 models. Enevate investors include carmakers Renault, Nissan, and Mitsubishi

There has been another exciting development in EV batteries for a considerable period: solid-state batteries.

At some point, the development of solid-state batteries, where electrons flow through a solid material instead of a liquid or gel, will lead to new electric vehicles. These vehicles can go much farther on the charge, and battery-storage systems can hold more energy while taking up less space.

Solid-state batteries can hold much more energy per unit of mass than today's lithium-ion batteries, which means an EV could go for much longer before needing to be recharged. Even though the electrolyte is solid, it is porous on a microscopic level, allowing electrons to pass through it.

Researchers say their battery is low cost, easy to build, has a high degree of mechanical stability, and is chemically stable—four important market viability factors. Wow! Sounds too good to be true.

Solid-state batteries could be the batteries of choice sooner than you think.

Solid Power, a Colorado-based company developing solid-state EV batteries for partners including Ford and BMW, said it had completed the installation of a "pilot production line" capable of making about 300 battery cells per week. This setup signals that the technology is moving from the lab to the factory.

The gold rush is on!

The race for better electric car batteries is famous as the next gold rush. We are at the threshold of a battery revolution. Let's look at the ground-breaking research that may lead to an exciting new world of battery technology for tomorrow's electric cars.

***Structural Component Batteries*:** Research at the Chalmers University of Technology has been focusing on using new battery tech as a structural component of future electric cars. This technology could lead to very light vehicles in which body parts are the batteries. The positive is a lithium iron phosphate using carbon fiber as the negative electrode. These batteries would be extremely stiff and rigid for structural components.

***Carbon Nanotube Electrodes*:** NAWA Technologies has designed and patented an Ultra-Fast Carbon Electrode that could completely change today's batteries as we know them. The technology utilizes a vertically aligned carbon nanotube that can boost battery power ten times over current battery packs. It can also increase energy storage by a factor of three and increase the lifecycle of a battery five times. NAWA says that charging time will be just five minutes to get to an 80 percent charge. EV production, as soon as in 2023, could use NAWA technology.

***Cobalt-Free Batteries*:** The University of Texas is working on a lithium-ion battery that doesn't use cobalt as a cathode. Instead, it uses 89 percent nickel as well as aluminum and manganese. The reason is that cobalt is rare and expensive. The team at the University of Texas says that their batteries also produce a more even distribution of ions.

A Chinese company called SVOLT manufactures cobalt-free batteries for the EV market. They claim a higher energy density, resulting in a range of up to 500 miles on a charge.

***Silicon Anode Batteries*:** Looking for a cure for unstable silicon in lithium-ion batteries, researchers at the University of Eastern Finland have developed a method to produce a hybrid anode that uses mesoporous silicon microparticles and carbon nanotubes. They hope to replace graphite as the anode and replace it with silicon, which has ten times the capacity. The researchers hope that this will improve battery performance. The silicone is made from barley husk ash and is, therefore, very earth friendly.

***Seawater Batteries*:** IBM Research has discovered new battery chemistry that is free of heavy metals and can outperform lithium-ion batteries. The extraction of materials from seawater helps in its creation. IBM says its batteries will be cheaper to make, will charge faster, and pack in higher energy density and power. The company is currently working with Mercedes-Benz to develop the technology.

***Sand Batteries*:** Researchers at the University of California Riverside are working on a battery technology that uses sand to create pure silicon to achieve three times better performance than current graphite-based lithium-ion batteries. This new pure silicon also advances the life span of batteries.

A battery start-up company called Silnano is bringing this technology to the market through funding from Daimler and BMW. It promises a 40 percent boost in battery performance shortly.

***Wi-Fi Powered Batteries*:** Imagine powering your car over Wi-Fi while you drive, and you'd never have to recharge your battery by plugging it into a charging station. While this technology is still a ways off, researchers have developed a radio wave harvesting antenna that is only several atoms thick and can be used to recharge future EVs over electromagnetic waves.

The concept involves incorporating the molybdenum disulfide rectenna so AC power can be downloaded from Wi-Fi and converted to DC power to directly recharge a battery or power an EV. Let's hope it doesn't fry your brain at the same time.

***Over-the-Air Ultrasound Charging*:** Another way to possibly transmit rechargeable power over the air is through ultrasound. A company called uBeam turns power into sound waves that can be beamed to your EV and then turned back into power. Currently, uBeam is experimenting with this technology to power smartphones and laptops, but who knows where this might lead? Beam me up, Scotty!

***Biological Semiconductors*:** Speaking of smartphone charging, a start-up company called StoreDot, which was born from Tel Aviv University's nanotech department, has developed a charger that uses biological semiconductors. They use organic peptide compounds, which are the building blocks of proteins. The result is a charger that can recharge your smartphone in just 60 seconds, and the organic compounds used are non-flammable for safer charging. StoreDot is currently building batteries for EVs that will charge in five minutes and offer a range of 300 miles.

***Batteries that Never Die*:** Scientists at the University of California are working on nanowire batteries that will never die. The gold nanowires are a thousand times thinner than human hair and sit in a gel of electrolytes to keep them from breaking down during recharging. They have been tested recharging over 200,000 times over three months and showed no sign of degradation.

***Zinc-Air Batteries*:** Researchers at Sydney University have found a way to make zinc-air batteries for much less than current methods. Zinc-air batteries are superior to lithium-ion batteries, as they cannot catch on fire. However, using costly components makes them an expensive affair. The University has found a way to use much cheaper alternatives to manufacture them. So, more affordable and safer batteries may soon be on the way.

***20x Faster Charging*:** Ryden dual carbon technology allows batteries to last longer and charge faster than lithium. The factories producing lithium batteries also help in the making of these batteries. Power Japan Plus says that the batteries are more sustainable, last longer, are environmentally friendly,

***500 Miles on a Charge*:** Graphenano is developing a Graphene battery, which will offer a range of 500 miles. The battery will recharge in just a few minutes. The company says its batteries will charge and discharge 33 times faster than lithium-ion batteries.

***1,100 Miles on a Charge!* :** An experimental car recently drove 1,100 miles on a single battery charge. This test was possible due to aluminum-air battery technology that uses oxygen from the air to fill the battery's cathode. This invention makes it much lighter than liquid-filled lithium-ion batteries giving the EV even greater range.

All this is crucial; therefore, much research and development are taking place.

The goal is to develop improved EV batteries that charge faster and last longer while switching to less expensive and more environmentally friendly materials. It’s only a matter of time before the incorporation of some or all of these developments is available, which will hasten the demise of ICE vehicles.

Why do batteries catch fire?

With the transition to electric transportation comes a new challenge: Vehicles with lithium-ion batteries can be dangerous when they catch fire.

The good news is; battery electric vehicle fires don't happen often.

Battery electric vehicles have just a .03% chance of igniting, compared to internal combustion engine vehicles' 1.5% chance. Hybrid electrics, which have both a high voltage battery and an internal combustion engine, have a 3.4% likelihood of vehicle fires, according to research by an insurance portal, AutoinsuranceEZ.

However, when fires occur, electric vehicles with lithium-ion batteries burn hotter and faster, requiring far more water to reach final extinguishment. And the batteries can re-ignite hours or even days after the fire control, leaving salvage yards, repair shops, and others at risk.

The qualities that make lithium-ion battery cells powerful enough to move a passenger vehicle can also make them vulnerable to igniting-- especially if battery cells within them are damaged or defective.

Causes of lithium-ion battery fires

Lithium-ion battery cells have electrodes placed close together, increasing the chances of a short, filling them with a flammable liquid electrolyte. This flammable liquid could get into a thermal runaway situation where it just starts boiling, resulting in a fire.

EVs include battery management systems to maintain the correct operating temperature for high-voltage batteries, and those systems control how fast batteries charge and discharge. Improvements to them and the battery cells themselves promise to make EVs safer.

An electric vehicle battery consists of a few parts. First, one can see a metal shell above, and the batteries and the cooling systems find a place for themselves below the cover. The inside of an electric vehicle battery is not somewhere you want things to get. This rupturing is the key to why most EV fires start and why they're so hard to get out.

If a cell is punctured, it will do something wrong, igniting or failing. Of course, if one cell fails and collapses, there can be a chain reaction. Cell after cell gets a puncture and burn, and it's often hard to tell how severe that can be. Moreover, it's difficult to tell when the risk has passed. Slow leaks can cause flare-ups hours or even days later.

So, it's not necessarily that electric vehicle batteries will magically catch fire again. Instead, the chain reaction we're talking about can often be slow, with cells rupturing and burning at different rates. That's invariably the most significant danger here.

Lithium-ion batteries in cars or electronic devices can catch fire if their manufacturing is improper or damaged. If the software that operates the battery has a design defect, it can also catch fire.

The major weakness of lithium-ion batteries in electric cars is the use of organic liquid electrolytes, which are volatile and flammable when operating at high temperatures. An external force, such as a crash, can also lead to chemical leakage.

Modern electrolytes involve dissolving a lithium salt in a liquid material composed primarily of linear and cyclic chain carbonates. Molecules that affect a carbon atom is attached to three oxygen atoms. These liquids are typically flammable and unstable when exposed to high temperatures.

If the battery gets exposed to excessive heat or penetration in the battery case, it leads to an internal short circuit.

Simply put, this short circuit causes excessive heat, which in turn causes a chemical reaction. That generates more heat, making the chemical reaction develop even faster, producing more heat — a vicious circle.

This process is called "thermal runaway" and can lead to a domino effect. Namely, the heat generated by a failed cell can move to the next cell, making it thermally unstable. In seconds, this chain reaction can destroy the entire pack and even cause explosions.

What can cause a thermal runaway?

1. Manufacturing defects or design flaws
2. Software issues
3. Physical damage during an accident
4. Extreme weather conditions
5. Use of poorly insulated chargers

Why do EV fires present a unique challenge?

There are four main reasons why EV fires need attention on time;

1. Due to their intensity, they require more water to put them out. According to the Confederation of Fire Protection Associations of Europe, firefighters need more than 60,000 liters of water and a flow rate of 1100 liters per minute to tackle an EV fire.
2. Firefighters also need to prevent the water from flowing into drains because of the toxins it picks up from the burning batteries.
3. During an EV fire, over 100 chemicals get generated from the battery, including toxic gases such as carbon monoxide and hydrogen cyanide, which are fatal to humans.
4. Electric vehicle fires can reignite minutes, hours, or even days after the initial event.

What can you do to avoid it?

While the burden falls mainly on manufacturers to ensure safe batteries, there are a few things you can do to enhance your EV's safety, such as :

- Park your electric car in the shade and away from flammable items
- Use only original and authentic chargers for specific battery types
- Charge your EV battery using grid power and avoid overcharging

- It is advisable to let the batteries cool down after use for an hour or so before recharging
- If you notice the battery is overheating while charging, cut off the electric supply immediately
- Clean, dry, and ventilated places are the best to keep the batteries. There should be no contact with corrosive substances. The battery should be away from heat sources
- Avoid keeping batteries empty or full. Keep your battery's state of charge between 20% and 80% whenever possible. Only charge fully for long trips
- You should also regularly get your vehicle's batteries checked and repaired by the automaker's authorized service centers.

And it goes without saying that if your car catches fire, you should immediately evacuate and not attempt to extinguish the fire yourself.

4

CHARGING INFRASTRUCTURE

"You and I come by road or rail, but economists travel on infrastructure."
—Margaret Thatcher

The efficient integration of EVs into a country's electricity system is required to reap the benefits of electric vehicles (EVs). The limited availability of charging infrastructure, including a lack of good business and financing models, is considered one of the biggest obstacles to the widespread adoption of EVs by customers, many of whom still suffer from 'range anxiety.'

From a service provider's viewpoint, this is a 'chicken and egg problem for market operators who are wary of investing in the needed infrastructure without having any vision of its actual utilization because of the uncertain adoption rates of EVs, as well as potential interoperability issues.

Generally speaking, we can distinguish between four categories of charging infrastructure:

1. Public charging station in public domain (e.g., roadside/sidewalk)
2. Publicly accessible charging stations in the private domain (e.g., commercial areas such as shopping malls)
3. Semi-public charging stations in the public or private domain (e.g., car-sharing CS, hotels, or business parking for visitors and customers)
4. Privately accessible charging station (e.g., home or office locations)

Although electricity is ubiquitous, its transmission, distribution, and retail charging options for electric vehicles are not. Developing a robust charging infrastructure network is essential for a large-scale transition to electro-mobility. Such infrastructure would provide more charging options for drivers and promote awareness and range confidence for prospective electric vehicle owners.

Several automakers have begun to directly build their charging infrastructure networks, while others have engaged in partnerships with other automakers and charging providers. Many governments have created programs to encourage the construction of charging infrastructure through incentives, regulations, and teamwork.

Electric Vehicle Charging Technology

As electric vehicle charging technology advances, several standards and guidelines have become widely accepted. This section briefly overviews setting infrastructure technology, bars, and terminology.

They are charging Speeds. Charging power, which determines the time required to charge a vehicle, can vary by magnitude across charge points, as shown in the table below. A small household outlet may charge as slowly as 1.2 kW, while the most advanced rapid charging stations can charge up to 350 kW. Charging infrastructure is broadly broken into three categories based on speed: Level 1, Level 2, and direct current (DC) fast charging (sometimes referred to as Level 3).

Level 1 charging will take a long time to fully charge your car, whereas Level 2 charging can charge an electric car battery in 4-10hours, and level 3 chargers can charge an EV in as little as 20 minutes. The table on the next page summarizes the typical power output, setting time, and locations for PHEVs and BEVs for the different charger types.

Many electric vehicles are limited in the maximum charging power they can accept because of restrictions in their ability to convert AC power from the grid to DC power that charges the batteries. For example, the Chevrolet Volt, a plug-in hybrid vehicle (PHEV), is limited to 3.6 kW, and the Nissan Leaf, a battery electric vehicle (BEV), is limited to 6.6 kW. Furthermore, some electric vehicle models, including most PHEVs, are incapable of DC fast charging.

They are charging Connector Standards. Depending on the region and speed of charging, the type of plug and socket used for charging electric vehicles may vary. Although these plug types are generally well defined and each works well for its specific application, the variety of standards may lead to confusion among drivers and hesitation from the industry.

In North America and Japan, most electric vehicles use the SAE J1772 connector, which contains five pins and a mechanical lock. In Europe, Level 2 charging uses the Type 2 or Mennekes connector, which has seven pins and takes advantage of the three-phase alternating current grid. China also requires a variant of the Type 2 plug known as GB/T. Tesla uses a proprietary connector for its vehicles in North America, although adapters to SAE J1772 are available. In Europe and Asia, Tesla vehicles have a Type 2 plug.

Charging Connector Types

	Level 1	Level 2	DC Fast Charging
Connector Type	J1772 connector 	J1772 connector 	CCS connector  CHAdeMO connector  Tesla connector 
Typical Power Output	1 kW	7 kW - 19 kW	50 - 350 kW
Estimated PHEV Charge Time from Empty[1]	5 - 6 hours	1 - 2 hours	N/A
Estimated BEV Charge Time from Empty[2]	40 - 50 hours	4 - 10 hours	20 minutes - 1 hour[3]
Typical Locations	Home	Home, Workplace, and Public	Public

1. Assuming an 8-kWh battery, most plug-in hybrids do not work with fast chargers.

2. Assuming a 60-kWh battery.

3. To 80 percent charge. Charging speed slows as the battery gets closer to full to prevent damage to the battery. Therefore, it is more cost- and time-efficient for EV drivers to use direct current (DC) fast charging until the battery reaches 80 percent and then continue on their trip. It can take about as long to set the last 10 percent of an EV battery as the first 90 percent.

For DC fast charging, connector types vary by the automaker and region. Nissan and Mitsubishi created and promoted the

CHAdeMO (short for Charge de Move) fast charging standard in 2011. This type still exists on electric vehicles produced by Nissan, Mitsubishi, Kia, Citroën, and Peugeot. In contrast, several automakers from the United States and Europe have advocated for the Combined Charging System (CCS), which uses the SAE J1172 or Mennekes AC plugs along with two additional DC pins for fast charging. BMW, Daimler, Ford, Fiat Chrysler, General Motors, Honda, Hyundai, and Volkswagen have adopted this standard.

Whereas CCS uses the same receptacle on the car as a Level 2 charger, CHAdeMO requires a separate port. As in the case of Level 2 charging, Tesla uses its proprietary plug for its DC Supercharger stations in the United States, although the company also makes Tesla-to-CHAdeMO adapters. China has recently mandated using a new standard (GB/T 20234.3-2015) for all new vehicles and fast charging infrastructure, and Tesla vehicles sold in China will also use this standard.

How many charge points are needed?

There is no universal benchmark for the number of electric vehicles per public charge point. Electric vehicle owners in California more frequently have access to home and workplace charging, and one public charger per 25 to 30 electric vehicles is typical. Private parking and charging are relatively rare in the Netherlands; one public charger per 2 to 7 electric vehicles is standard. This ratio ranges from 3 to 6 in major markets in China, and these cities typically had the highest percentages of rapid charging.

Harrison and Thiel (2017) modeled several factors' impact on the European electric vehicle market share, including charging infrastructure. This model calculated the utility and respective market share of different powertrain types, using feedback loops to capture realistic decision-making patterns by drivers, manufacturers, charging infrastructure providers, and policymakers. The model also assessed the profitability of charging stations under various scenarios and considered subsidies and government targets for charging infrastructure.

The authors found that the private market could profitably support 95% of public charging stations, up to a ratio of 25 electric vehicles per charge point. They also found that the electric vehicle market share increases as the electric vehicle/charge point ratio decreases from 25 to 5 electric cars per charge point. Charging infrastructure availability also appears to have the most substantial impact on uptake once electric vehicle stock share exceeds 5%.

In 2021, sales more than doubled to bring the total fleet of electric cars to about 16.5 million, a tripling relative to the stock in 2018. Meanwhile, publicly available charging points also tripled to about 1.8 million.

Who should be responsible for the charging infrastructure?

Multifaceted and collaborative approaches have been most successful in promoting early charging infrastructure buildout. Governments at the local, regional, and national levels around the world have used varied strategies to promote public and private charging infrastructure.

Successful programs have transparently engaged many stakeholders through driver feedback on charger deployment, implementation of innovative charging systems, distribution of funding to local governments, creation of public/private partnerships, and consultation with electric utilities. To address changing needs in this growing market, governments create and fund programs that target complex market segments, such as curbside charging stations, multi-unit dwellings, and intercity fast charging.

Publicly accessible chargers worldwide approached 1.8 million charging points in 2021, of which a third were fast chargers. The government installed nearly 500,000 chargers in 2021, more than the total number of public chargers available in 2017.

As in previous years, China is the global leader in the number of publicly available chargers, and it counts for about 85% of the world's fast chargers and 55% of slow chargers.

Let us look at some significant government programs promoting charging infrastructure in selected markets and highlight emerging best practices.

USA

Although electric vehicle (EV) sales in the United States have climbed by more than 40 percent since 2016, nearly half of US consumers say that battery or charging issues are their top concerns about buying EVs. It's no stretch to say that the nation's limited network of charging stations probably discourages many prospective buyers.

In response, the Bipartisan Infrastructure Law (BIL) provides $7.5 billion to develop the country's EV-charging infrastructure. The goal is to install 500,000 chargers—publicly accessible charging stations compatible with all vehicles and technologies—nationwide by 2030. However, even the addition of half a million public chargers could be far from enough. In a scenario where half of all vehicles sold are zero-emission vehicles (ZEVs) by 2030—in line with federal targets—McKinsey estimates that America would require 1.2 million public EV chargers and 28 million private EV chargers by that year. The country would need almost 20 times more chargers than it has now.

The Alternative Fuels Data Centre lists almost 50 000 EV charging stations in the United States. Of these, 93% are publicly accessible, and 17% are on non-urban roads (including highways and other arterials). A disproportionate share of direct current (DC) fast chargers are public (99%) and located on highways (25%), reflecting the faster-charging needs at these locations.

As the number of EVs on the road increases, annual demand for electricity to charge them will surge from 11 billion kilowatt-hours (kWh) to 230 billion kWh in 2030, according to McKinsey's scenario-based modeling. The demand estimate for 2030 represents approximately 5 percent of the current total electricity demand in the United States. The modeling indicates a requirement of nearly 30 million chargers to deliver the electricity demand in that year.

The government or the companies will install these chargers at residences—on-the-go locations and destinations where vehicles get parked for extended periods will have 1.2 million public chargers. McKinsey estimates that this public charging infrastructure's hardware, planning, and installation cost will exceed $35 billion by 2030.

Europe

Europe will have 130 million electric vehicles on the road by 2035, according to a joint report from Ernst & Young and the electricity industry trade association, Eurelectric.

The report's projections show Europe's EV fleet growing from its current base of less than 5 million to 65 million by 2030, doubling over the following five years. EY estimates that the continent will need 65 million chargers to fuel these cars, trucks, and buses, with 85% of plugs installed at homes, and this will require a $134 billion investment in infrastructure.

Europe's rapid adoption of electric vehicles presents two large tasks for utility providers. The first is building a network of 9 million out-of-home chargers along roadways, workplaces, and fleet-charging hubs. There are about 445,000 public connectors installed across Europe, according to the latest tally from BloombergNEF.

In addition to overseeing the installation of millions of chargers, Europe's utility industry will need to manage an increased load on the grid. Along highway corridors, where drivers expect fast charging on demand, EVs could increase peak loads by 90%. Managing these surges will require on-site solar and energy storage systems at charging stations.

The EU has more than 330,000 publicly accessible charging points, which is growing, but their uneven deployment means to travel across the EU in electric vehicles is not easy. Just three countries – Germany, France, and the Netherlands – account for 69% of all charging points across the EU, while 10 European countries do not have a single charger per 100 km of road. There are disparities within countries, too, with cities much better covered than rural areas, especially in Eastern Europe.

The European Commission has a target of hitting 1 million charging points by 2025. It is an estimation that Europe requires 150,000 new electric points yearly and almost 3,000 a week to close the gap.

China

China has so far built more than 1 million public charging points. Still, over 70 percent of them are in regions such as Guangdong and Shanghai, which have a relatively high number of piles. At the same time, many counties and villages are still short of charging infrastructure.

The charging posts aren't always where they're most needed, with just 0.5% of all points located on highways versus about 3% of gas stations. The construction of charging infrastructure hasn't always followed demand, meaning you may be able to find a charge point in some remote corner of Tibet but have to queue in Shanghai, where about 30% of all cars are electric or new energy-powered.

According to a document released by China's National Development and Reform Commission (NDRC) and nine other government departments, the country will move faster to implement an innovative, efficient, well-balanced, and moderately advanced network of battery charging facilities. All fixed parking spaces in new residential districts will have room for charging facilities.

By 2025, no less than 60% of expressway service areas in the country will have rapid charging stations, and no less than 80% of national ecological civilization pilot zones and critical air pollution prevention and control areas will have the same facilities.

India

The charging requirement of EVs varies with category (2W, 3W, or 4W) – depending on the location, power requirement, and rating of the said vehicle. In India, 'conductive charging' is the mainstream technology in use. The Electric Vehicle Supply Equipment (EVSE) safely charges the electric car by first accessing power from the local electricity supply and then running it through a wired connection and an in-built control system.

Standard power AC charging ranges from 3kW to 22kW and is adequate for e-2Ws, e-3Ws, e-4Ws, and other light commercial vehicles (LCVs) up to 1 ton, which is currently prevalent across India. High-power DC chargers, which are faster but require a higher electricity supply with additional infrastructure, are gradually integrated into the network and will be more widespread soon.

The initial focus has been on nine megacities with populations of over 4 million – Delhi, Mumbai, Kolkata, Chennai, Bengaluru, Hyderabad, Pune, Ahmedabad, and Surat. There are currently over 1,640 operational public EV charging stations in India (as of February 2022), of which more than 940 are in these cities.

Demonstrating the extent of change that has occurred so far, data shows that the government installed 678 additional EV charging stations in nine cities between October 2021 and January 2022. It is 2.5 times that of the earlier figures during the same period, powering about 1.8 lakh new electric vehicles.

India needs to set up 46,000 electric vehicle charging stations by 2030 to reach the global benchmark, suggests a white paper released by Alvarez and Marsal, a global professional services firm. The EV charger ratio is 6 for China and the Netherlands, 19 for the U.S., and 135 for India. That means there is one charger per 135 EVs in India compared to 6 in China.

Battery Swapping

A battery swapping (or switching) station is where a vehicle's discharged battery or battery pack can be immediately swapped for a fully charged one, eliminating the delay in waiting for the vehicle's battery to charge.

Battery swapping is common in warehouses using electric forklift trucks. The concept of an exchangeable battery service came to light in around 1896 to overcome the limited operating range of electric cars and trucks. It was first put into practice between 1910 and 1924, by Hartford Electric Light Company, through the GeVeCo battery service and was initially available for electric trucks. The vehicle owner purchased the vehicle, without a battery, from General Vehicle Company (GeVeCo), part-owned by General Electric. The electricity was purchased from Hartford Electric using an exchangeable battery. Vehicles and batteries are modified to facilitate a fast battery exchange. The owner paid a variable per-mile charge and a monthly service fee to cover the maintenance and storage of the truck. During the service, the vehicles covered more than 6 million miles.

Tesla Motors and Mitsubishi Heavy Industries are working on integrating battery switch technology with their electric vehicles to extend driving range. The driver does not need to get out of the car at the battery switch station to swap his battery, and this swap requires an electric car designed for the "easy swap" of batteries. However, electric vehicle manufacturers working on battery switch technology have not standardized battery access, attachment, dimension, location, or type.

Battery swapping has the following benefits:

- Fast battery swapping under five minutes.
- An unlimited driving range where there are battery switch stations available.
- The driver does not have to get out of the car while changing the battery
- The driver does not own the battery in the car, transferring costs, battery life, maintenance, capital cost, quality, technology, and warranty to the battery switch station company.
- Contracting a battery switch company could subsidize the electric vehicle at a price lower than equivalent petrol cars.

The spare batteries at swap stations could participate in the vehicle-to-grid storage

Power Supply and Grid Effects

Electric vehicle charging has the potential to use vast amounts of power. Although it currently does not pose any substantial risk to the grid, this is an issue that authorities should consider as the market grows.

A frequently cited concern is the worsening of evening peak power demand, both locally and regionally, due to many drivers plugging in after arriving home from work.

Many governments are engaged with working group activities, pilot projects, and policy processes to incorporate greater use of intelligent charging practices, including controlled charging and demand response. Even more straightforward solutions, such as using in-vehicle timers to take advantage of time-of-use rates, could help minimize stress on the electrical grid while saving consumers money.

Another concern in some areas is the effect of DC fast charging on local distribution infrastructure. These fast charging stations use very high amounts of power for short periods, meaning that more expensive upgrades are essential with a relatively low use rate. This problem could intensify as technology improves.

These usage patterns and the potential for infrastructure upgrades often cause charging sites to incur high demand charges, a component of electricity rates based on the highest capacity. For fast charging stations, which use a lot of power but are rarely in use by drivers, demand charges can account for 90% of operating costs, which leads to higher rates for drivers. Utilities, regulators, and research groups are developing alternative rate structures for the workplace and public charging infrastructure, an essential step in improving the commercial case for electric vehicle charging.

Vehicle-Grid Integration (VGI)

EV charging creates a reciprocal relationship between battery-powered vehicles and the power grid in a way that produces mutual benefits.

Vehicle-Grid Integration (VGI) is a way to develop solutions that enable electric vehicles (EV) to provide grid services while still meeting consumer driving needs. To that end, EVs must have capabilities to manage to charge or support two-way interaction between vehicles and the grid.

Managed charging refers to the technical capability to modulate the electric charging of the vehicle through delay, throttling to draw more or less electricity, or switching load on or off. Two-way interaction refers to the controlled absorption and discharge of power between the grid and a vehicle battery or a building and a vehicle battery.

VGI works through technology tools and products that provide dependable vehicle charging services to EV owners and potentially additional revenue opportunities while reducing risks and creating cost savings opportunities for grid operators. Such tools might include technologies such as inverters, controls, chargers, or programs and products, such as time-of-use tariffs or bundled charging packages.

An EV can provide the functionality of both *load* (while its battery is charging) and *generation* (by discharging stored electricity from the battery to the grid). The magnitude of electricity demanded or supplied can be easily controlled and moderated. A PEV battery can require EVSE (Electric Vehicle Supply Equipment, aka EV charging stations) to extend the range of instantaneous load. The auto manufacturers provide charging options of up to 120 kW (and potentially higher capacity in the future) to accommodate a variety of driver time constraints. Conversely, for generation] because EV batteries supply power to electric motors, they must withstand extreme and instantaneous discharge cycles (to accelerate a vehicle to cruising speeds) and recharge process (to recapture energy while slowing it down).

Many EVs and specific electric vehicle charging stations are equipped with on-board timers or remotely controlled switches capable of starting, stopping, throttling, or delaying charging. These charging stations give drivers the ability to schedule charging remotely.

The cost of a battery electric vehicle's energy storage system accounts for 30% to 60% of the manufacturer's suggested retail price and is its single most expensive component. While considering the expense of a battery, the average vehicle spends only 4% of the day driving. While vehicles are idle about 96% of the time, an electric car must charge only about 10% of the time. The charging process suggests considerable flexibility for drivers to shift charging to different times of the day to minimize their costs and maximize benefits to the grid.

Drivers will connect their EVs to an EVSE when the vehicle is at home, but the energy required for charging depends on their driving needs. EVs' overnight energy consumption relies on the miles driven during the previous day. The ability of an EV to provide load (and potentially generation) without materially affecting its primary use for mobility is easy to manage and control. The control process demonstrates an opportunity for regulators to work with customers, equipment providers, and grid operators to use electric vehicles as grid resources.

5

CLEAN ENERGY

"Our future is green energy, sustainability, renewable energy."
- Arnold Schwarzenegger

As per IEA's 2022 report, the global EV fleet in 2021 consumed about 55 terawatt-hours (TWh) of electricity (about 10% of which was for electric two/three-wheelers in China). Electricity demand from EVs accounts for less than half a percent of the current total final electricity consumption worldwide.

The expectation is for the rapid growth of the EV fleet. IEA estimates that EV demand will reach between 780 TWh to over 1,100 TWh in 2030, accounting for at least 2% of global electricity consumption.

EVs enable a net reduction in carbon emission on a well-to-wheel basis, and this information is a fact now. In 2021, EVs helped a net decrease of 40 million tons of carbon dioxide-equivalent (Mt CO2-eq) on a well-to-wheel basis, roughly equivalent to Finland's entire energy sector emissions. China's EVs led to the most significant savings despite its carbon-intensive electricity mix. Almost 45% of the emissions reductions come from the electrification of two/three-wheelers. And the net GHG (greenhouse gas) benefit of EVs will increase over time as electricity decarbonizes.

Essentially, unless the electricity that powers EVs is clean, EVs can never be a fully green option. With the vast number of EVs to come online in the coming years, both users and utilities must find a way to charge them with renewable energy sources.

EVs need a clean energy grid

Because they have an electric motor instead of an internal combustion engine, EVs emit no exhaust from a tailpipe, which means that, unlike traditional vehicles, they don't pump carbon dioxide, ozone, and particulate pollution into the air we breathe.

No exhaust is important because transport accounts for around one-fifth of global emissions, accounting for three-quarters of road travel. Most of this comes from passenger vehicles – cars and buses – which contribute 45.1%, and the other 29.4% comes from trucks carrying freight.

Moreover, this number will keep increasing as population growth, and demographic shifts drive ever more demand for road travel – not to mention the rise in e-commerce pushing up the need for freight and last-mile delivery.

EVs need anywhere between 24 and 50 kWh of electricity to travel 100 miles, and this electricity comes from the grid. A US Department of Energy study shows that increased electrification will boost national consumption by as much as 38% by 2050 because of electric vehicles. In some cases, EVs could result in substantial greenhouse gas (GHG) emissions or even help extend the life of fossil fuels if charged primarily with fossil fuel-generated power.

A recent study by China's Tsinghua University found that EVs charged in China – where most electricity comes from coal-fired power plants – contribute two to five times as much particulate matter and chemicals versus gas-engine cars. How sad!

The finding defeats the very purpose of EVs –having a cleaner environment and reducing carbon emissions. This emission process creates a significant opportunity to position EVs as one of the biggest buyers of renewable electricity globally

The model already exists: corporate procurement of renewable energy via bilateral power purchase agreements (PPAs) has created significant voluntary demand for new renewable energy projects worldwide. In 2020, corporations purchased a record 23.7 GW of clean energy, up from 20.1GW in 2019 and 13.6GW in 2018. New research published by Bloomberg NEF (BNEF) came despite the disruption caused by the Covid-19 pandemic and the ensuing global recession.

Through PPAs, EV original equipment manufacturers (OEMs), charge point operators, and electric mobility service providers. The growing number of companies committing to switching their vehicle fleets to EVs can develop seamless green solutions for the future and facilitate the development of new renewable energy projects. This process will, in turn, bring the world closer to meeting the Paris Agreement targets.

It isn't only the juice that powers the vehicles' batteries that's important. Half of the lifecycle emissions from the lithium batteries in EVs come from the electricity used to assemble and manufacture them, which means that the electricity mix at OEM facilities is also a vital part of the equation. A recent study by IVL, the Swedish environmental institute, found that lithium batteries produced in regions with a zero-carbon grid had emissions of 61kg of CO2 equivalent per kWh of battery capacity (CO2e/kWh). This figure more than doubles – to 146 kilograms – when the electricity used in battery manufacture comes from fossil fuel.

Therefore, the climate benefit of EVs doesn't depend on how green the electricity used to charge their battery is but also on the carbon intensity of the electricity used to make that battery – creating yet another imperative for EV manufacturers to switch to renewable energy.

The rise in the adoption of EVs can also drive the growth of renewable energy in other ways. Private cars spend 95% of their time parked, and energy planners are looking at ways to utilize this slow time to solve one of the biggest problems for scaling up renewable grids: stability. Intelligent charging, which charges vehicles and supports the grid, unlocks a virtuous circle in which renewable energy makes transport cleaner, and EVs help larger shares of renewables.

The technology to make this happen is still in its infancy – so far, the Nissan Leaf is the only mass-production EV on the market that enables vehicle-to-grid (V2G) charging. However, more OEMs are starting to consider this capability: for example, Hyundai, Kia, and Lucid all plan to include it in future vehicles.

The ultimate promise of electric vehicles is a cleaner environment. The strain on the power grid requires additional investment in non-renewable resources to deliver the energy.

That's why renewable energy for EV charging is so critical. Using renewable energy to power the EV charging infrastructure simultaneously eliminates the strain on the grid and the environment, especially when paired with smart EV charging.

Smart EV charging increases the ability to control the charging process, distributing power between charging sites and campus requirements and among the individual charge points within a charging station.

With good planning and the proper infrastructure, EVs can reduce emissions, replace polluting vehicles, and boost the rollout of renewable energy infrastructure. When parked up and plugged in, they act as battery banks, stabilizing electric grids powered by renewable solar energy.

IEA projections show that global electricity demand will grow by more than a third by 2040, mainly due to the adoption of EVs, which will take the electricity demand for transport from almost nothing to 4,000 TWh a year. This demand will increase electricity's share in total final energy consumption from 19% in 2018 to 31% in 2040, overtaking oil and leaving coal in the dust.

EVs are here to stay, but for them to be genuinely a green option for the future of transportation, we mustn't miss the chance to link them with renewable energy.

Green Hydrogen

You may have heard of the term 'green hydrogen' as this new, exciting 'clean' energy technology. But what exactly is green hydrogen, and how will it help clean our environment?

Tremendous strides led to producing clean energy from solar, wind, and hydropower, which is considered inexhaustible and sustainable. One more source of zero-carbon fuel holds excellent potential – hydrogen, and its creation does not involve carbon.

Hydrogen is first in the periodic table with atomic number 1. It is light, can be stored, and does not generate pollutant emissions. With these characteristics, it is a perfect candidate for fuel. However, hydrogen is not a primary energy source but an energy vector; it requires a chemical process to produce it. You have probably heard that hydrogen is a renewable fuel, but this is not always true. Hydrogen is only renewable if the method used in its extraction is also renewable.

No matter how the hydrogen creation occurs, it ends up with the same carbon-free molecule. However, the pathways to produce it are very diverse, and so are the emissions of greenhouse gases like carbon dioxide (CO2) and methane (CH4).

Because pure hydrogen does not occur naturally on Earth in large quantities, it usually requires a primary energy input to be produced on an industrial scale. One can get Hydrogen after extracting fossil fuels and biomass from water or a mix of both. Natural gas is currently the primary source of hydrogen production, accounting for around three-quarters of the annual global dedicated hydrogen production of approximately 70 million tonnes. This production accounts for about 6% of international natural gas use. Coal is closely following gas due to its dominant role in China and a small fraction of the use of oil and electricity.

Depending on the process used to produce it, a spectrum of colors describes the type of hydrogen.

Green hydrogen: Splitting water into hydrogen and oxygen in a process called electrolysis of water, using renewable electricity, creates green hydrogen. The crucial thing is renewable energy, so it generates no polluting emissions into the atmosphere. Green hydrogen is, therefore, the cleanest and most sustainable hydrogen

Grey hydrogen is traditionally a creation from methane (CH4). Split with steam into CO2 – the main culprit for climate change – and H2, hydrogen. Besides natural gas, grey hydrogen increasingly comes from coal, with significantly higher CO2 emissions per unit of hydrogen produced, often called brown or black hydrogen instead of grey. It is currently the cheapest hydrogen to make and the most common to use.

Blue hydrogen follows the same process as grey, with other technologies necessary to capture the CO2 produced when hydrogen is split from methane (or coal) and stored long-term. It is not one color but an extensive gradation; 100% of the CO2 is not in use, and all means of keeping it are equally effective in the long term. The impact of hydrogen production on the climate is reduced significantly by capturing a large part of the CO2.

There are technologies (i.e., methane pyrolysis) that hold a promise for high capture rates (90-95%) and adequate long-term storage of the CO2 in solid form, potentially so much better than blue that they deserve their color in the "hydrogen taxonomy rainbow," turquoise hydrogen. However, methane pyrolysis is still in the pilot stage. Green hydrogen is rapidly scaling up based on two key technologies - renewable power (particularly from solar PV and wind, but not only) and electrolysis.

Unlike renewable power, the cheapest source of electricity in most countries and regions today, electrolysis for green hydrogen production needs to significantly scale up and reduce its cost by at least three times over the next decade or two. However, unlike CCS and methane pyrolysis, electrolysis is commercially available and can be procured from multiple international suppliers.

With applications as a clean fuel, energy source, and more, green hydrogen's potential to support decarbonization efforts reaches far and wide. Where green hydrogen shines, however, is in its potential as an energy storage medium.

Green hydrogen, 100-percent renewable energy, is made through electrolysis. It offers a decarbonized energy storage solution for excess renewable energy, and it can be easily stored, transported, and converted back into power.

While traditional battery storage options generally offer up to eight hours of storage, hydrogen storage offers a much longer duration, allowing for the seasonal load-shifting potential that will increase the reliability of renewables.

Hydrogen surpasses the physical limitations of current mineral-based battery technologies to provide essentially infinite duration storage and backup power.

The time is right to tap into hydrogen's potential to play a crucial role in a clean, secure, affordable energy future. Pure hydrogen is enjoying unprecedented political and business momentum, with the number of policies and projects worldwide expanding rapidly. Now is the time to scale up technologies and bring down costs to allow hydrogen to become widely used.

Hydrogen is versatile. Today's technologies enable hydrogen to produce, store, move and use energy differently. Various fuels, including renewables, nuclear, natural gas, coal, and oil, can produce hydrogen. It can be transported as a gas by pipelines or in liquid form by ships, much like liquefied natural gas (LNG). The transformation converts it to electricity and methane to power homes, feed industry, and fuel for cars, trucks, ships, and planes.

It can potentially help with a variable output from renewables, like solar photovoltaics (PV) and wind, whose availability sometimes matches demand. Hydrogen is one of the leading options for storing energy from renewables and looks promising to be the lowest-cost option for storing electricity over days, weeks, or even months. Hydrogen and hydrogen-based fuels can transport energy from renewables over long distances – from regions with abundant solar and wind resources, such as Australia or Latin America, to energy-hungry cities thousands of kilometers away.

India's plans

India emitted about 2.9 billion metric tons of CO2 in 2021 and has committed to cutting total projected carbon emissions by 1 billion tons by 2030 and reducing carbon intensity of the nation's economy to less than 45% by 2030 from 2005. India has also committed to being net carbon zero by 2070. There will be an increased emphasis on green hydrogen to reduce carbon intensity. India's crude oil import has been growing at about 11% over the last two decades, and therefore there is economic sense in moving towards green hydrogen.

For energy-starved India, aiming for carbon neutrality by 2070, the path to energy security goes through a mix of oil, coal, blended fuels, natural gas, renewables, and electricity. India's $3.12 trillion economy currently needs 1,650 billion units (BU) of power, made from nearly 400 GW of capacity. Of this, green electricity is only 17%. If the economy touches $5-7 trillion in the next decade, it will need at least 3,000-4,000 GW. Further, at the current rate, the energy import bill will triple by 2040. The only way out of these massive challenges is by tapping as many green and locally available energy sources as possible.

New Delhi-based climate and energy research firm, Council for Energy, Environment and Water Research (CEEWR), has estimated that net zero emissions by 2070 will require 5,630 GW solar capacity, a 99% reduction in coal use between 2040 and 2060, and a 90% fall in crude oil consumption between 2050 and 2070. Green hydrogen should meet 19% of the industry's needs by that time. The Energy and Resources Institute (TERI) estimates 23 MT hydrogen demand by 2050. India's current output is 6.7 MT, produced primarily from natural gas through steam-methane reforming.

As per the Power Ministry, India plans to manufacture five million tonnes of green hydrogen per annum by 2030, aiming to meet its climate targets and become a production and export hub for the fuel. India will set up separate manufacturing zones, waive interstate power transmission charges for 25 years and provide priority connectivity to electric grids to green hydrogen and ammonia producers in a bid to incentivize production. India also plans to provide federal financial support to set up electrolyzers, as it wants to make the use of green hydrogen mandatory for refineries and fertilizer plants.

Green Hydrogen is not available on a commercial scale in India, as it is still waiting for its production in the country. The country's wealthiest men, Mukesh Ambani and Gautam Adani, have announced plans to produce it. At the International Climate Summit 2021, Mukesh Ambani said, "India can set an even more aggressive target of under $1 per kg within a decade. This target will make India the first to achieve $1 per 1 kg in one decade – the 1-1-1 target for green hydrogen."

Mukesh Ambani's Reliance Industries Ltd. has kicked off its green initiatives with a slew of deals in recent months and plans for setting up four Giga factories in Jamnagar at an investment of ₹60,000 crores for doing solar photovoltaic modules, advanced energy storage batteries, electrolyzers, and fuel cells.

Gautam Adani plans to build the world's largest renewable energy company with 45 GW capacity by 2030. He is also aiming to become the largest hydrogen producer in the world. Over the next decade, the group plans to spend 80% of its capital on green businesses, including $20 billion in renewables, green component manufacturing, and enabling infrastructure.

India's oil marketing companies (OMCs) are also in the game. The leader, Indian Oil Corporation (IOC), is working on plans to convert a part of the grey hydrogen it produces to blue hydrogen.

Bharat Petroleum Corporation Ltd. (BPCL) plans to have a 20 MW electrolyzer at its Bina refinery in Madhya Pradesh for building India's largest green hydrogen plant. Hindustan Petroleum Corporation (HPCL) is setting up a 370 MT green hydrogen plant at the Vizag refinery.

Gas Authority of India Ltd (GAIL) will build one of India's most extensive proton exchange membrane (PEM) electrolyzers at Guna in Madhya Pradesh to produce green hydrogen by the end of 2023.

A strategic partnership to develop a 1.3GW green hydrogen project powered by 24/7 renewable energy has been agreed upon between state-owned fossil-fuels company Oil & Natural Gas Corporation (ONGC) and Greenko, one of the world's largest renewables developers.

6

ECONOMICS OF ELECTRIC VEHICLES

"The tipping point is that magic moment when an idea, trend, or social behavior crosses a threshold, tips, and spreads like wildfire."
—Malcolm Gladwell

If an electric car is available at the same price as an ICE car, and assuming there are no 'range anxiety' or charging issues, what would you buy? It's a no-brainer–the electric vehicle would win hands down. Not only because of vastly lower running and maintenance costs but also because of a quieter ride with zero tailpipe emissions.

Fortunately for ICE cars, this is not the case today. EVs are far more expensive than their fossil-fuel counterparts, and that's the main barrier to the adoption of EVs.

The total cost of ownership (TCO)

The price of electric vehicles is declining, but they still look expensive compared with many gasoline-powered cars. That comparison, though, is misleading. Once you figure out the cost of ownership, especially maintenance and fuel, electric vehicles are a bargain despite the upfront price.

For an EV, assume higher upfront procurement costs but reap the cost savings during operation. As long as the vehicles aren't entirely underutilized, EVs typically have a lower total cost of ownership than vehicles powered by fossil fuels.

There are mainly two ways that EVs reduce operating costs:

1. ***Fuel Savings*** – EVs consume cheaper electricity instead of expensive petrol or diesel, and the savings on fuel are pretty dramatic. And the more the oil prices rise, the more savings you make on EVs.
2. ***Reduced Maintenance and Repairs*** – There are far fewer moving parts in an EV, making maintenance much less expensive than in ICE vehicles. Savings costs include oil changes, brake systems, engine and transmission repairs, filters, fan belts, timing belts, etc.

There could be savings from government incentives like tax rebates, subsidies, no toll tax, and no parking charges, which would add to the attractiveness of EVs.

As I write this, the price of Brent Crude Oil is up by $90 per barrel. In major cities of India, petrol prices have crossed Rs. 100 per liter. This scenario forces people to look for alternatives to save fuel costs. The question many people are asking is – do electric cars save money?

Considering that electric vehicles tend to be more expensive than gas-fuelled cars and that electricity has its costs, is it cheaper to go electric? The short answer is yes — although it also depends on your driving habits, where you live, and the type of vehicle you buy.

Economy Comparison: USA Example

The average transaction price for an electric vehicle (EV) is $56,437, according to Kelley Blue Book — roughly $10,000 higher than the overall industry average of $46,329, which includes gas and EVs. In terms of pricing, an EV is equivalent to an entry-level luxury car.

Many drivers install what's known as "Level 2" chargers in their homes to enjoy a long battery life and save time on charging. The cost of these chargers is around$2,000, including the installation. According to JD Power- With a Level 2 charger, it will take less than eight hours to charge your vehicle,

The surveys show that the price gap between EVs and gas-fuelled vehicles will decline in the next decade, depending on continued battery technology improvements, which could result in cheaper production costs. In the meantime, customers can offset some of the premium paid for EVs through tax credits, which come with caveats.

While EVs usually have higher upfront purchase prices, owners can save a lot on operating expenses. A 2020Consumer Reports study found that EV owners, on average, spend 60% less on fuel compared to internal combustion engine vehicles.

According to a study from the U.S. Department of Energy's National Renewable Energy Laboratory and the Idaho National Laboratory, electric car owners can save as much as $14,500 on fuel costs by owning an electric car for 15 years. That's almost $1000 in savings for every year of driving. Considering this study looked at 2020 gas prices, it's reasonable to assume that the savings are even more substantial today.

However, driving an electric vehicle isn't all worry-free from a money standpoint. That's because electricity isn't free, even if you're not paying for it at the pump.

Charging costs for an electric vehicle depend on many factors: where you charge (home or publicly), the time of day, electricity costs in your state, the size of the car battery, and more.

And since EVs have fewer parts than gas-fuelled cars (there's no oil to change, no spark plugs to replace), they also tend to have lower maintenance costs. A recent report by the analytics firm We Predict reveals that after 36 months on the road, service costs were 31% lower for electric cars and light trucks compared to similar gasoline-based vehicles.

U.S. Department of Energy report shows that after 15 years, electric cars generally cost less than similar gas-only models when you factor in the price, maintenance, financing, repairs, the federal tax break, and fuel costs. The electric version of a small SUV costs $0.4508 per mile, $0.0219 less than the $0.4727 per mile rate you get with a similar gas-based model.

Based on the average lifespan of a car — 200,000 miles, according toCar and Driver, the cost of a gas-fuelled vehicle would be $94,540, while a similar EV would be $90,160 for an equal EV, a difference of $4,380.

Here are the significant factors to consider while calculating the economics of an EV :

Purchase price. The sale price for an EV, on average, was $60,054 in February. That compared to $45,596 on average for all new vehicles, including electric ones, according to data from Edmonds. Compared to gas models, EVs typically cost 10 to 15 percent more. The price gap between EVs and conventional automobiles will narrow as battery costs drop and lower-priced models, some below $30,000, become available. We predict price parity between comparable models by the middle of the decade.

Energy costs. Here's where EVs have a clear and growing advantage. For example, the new Volkswagen ID.4 SUV, which starts at about $40,000, gets 250 miles per charge. With the average residential customer paying about 14 cents per kilowatt, it costs roughly $11 to charge the battery fully. A comparable VW Tiguan SUV, which starts at about $26,000 and gets 26 mpg, would cost about $38 to fill up at $4 a gallon. If you drive 12,000 miles a year, you could expect to spend around $550 to power the ID.4, compared to $1,900 for the Tiguan.

Size. Energy efficiency varies widely; the more extensive the vehicle, the more energy it will need. This energy requirement is also valid for gas models. So, while the Ford F-150 Lightning pickup will drain a battery faster than a compact Tesla Model 3 sedan, it will also save more per mile compared to a conventional Ford F-150, particularly during high gas prices.

Home chargers. More home chargers are coming to the market, and prices are plunging. Expect to spend around $500 to $700 for the charger. But the higher cost may be wiring it up, depending on the age and construction of your home or apartment building and your current electric service.

Public chargers: Expect to pay at least twice as much to use a high-speed charger at a gas station or charging station — if you can find one — compared to charging at home. For instance, the tab for Volkswagen ID.4 would be around $20 to $25. Currently, most public chargers are slow “Level 2” chargers, which can take an hour to add just 10 to 15 miles in range. However, the network of fast chargers is multiplying, and some of the newest EVs vehicles can add 200 miles of range in less than 20 minutes.

Maintenance and repairs. EVs don’t require tune-ups, oil changes, and some other types of routine maintenance. An EV, on average, during the first year, needs $254 in service and repairs versus $172 for a conventional model, according to We Predict research firm data. That’s mainly because EV issues are likely to be discovered early on, whereas traditional vehicles tend to develop problems later. After three years, EV owners, on average, spent $378 on service and repairs compared to $615 for gas models, according to We Predict.

Insurance. Rates vary by state and other factors, but EV owners typically pay a premium. One reason: EVs tend to cost more than their non-electric counterparts. Some analysts also believe insurance companies are cautious because they still have little experience with electric autos or data on which to base their actuarial tables.

Incentives. EV buyers get as much as $7,500 in federal tax credits, depending on the type of vehicle they buy. Some automakers, notably Tesla and General Motors, have crossed a sales threshold that phased out the federal incentives on their products. Congress is considering a measure that would restore their credits and possibly boost incentives as high as $12,500, but the chances of it passing are slim.

So, to buy or not to buy, if savings is the goal? The more you drive, the more you will save on energy. And the higher gas prices go, the faster you'll offset the additional upfront costs of an EV. Also, government incentives can sometimes make up for the higher price tag.

All things considered, if you drive at least 12,000 miles a year, have a home charger, and qualify for federal tax credits, experts say you have a good chance of being in the black within two or three years compared to buying and operating a comparable gas vehicle

Economy Comparison: India Example

India is still at a nascent stage of growth as far as electric cars are concerned. Tata Nexon EV leads the race with over 9,000 units sold in 2021.

Tata Nexon has both ICE and EV model variants, which, from the outside, look similar. The EV costs much more than its ICE siblings, so it helps to compare their ownership costs to determine if going for the EV version makes sense.

To make the comparison, I have taken the Nexon prices in Delhi as on August 2022. I have compared the Nexon EV Prime XZ Plus Dark Edition with Nexon XZA Plus Petrol AMT for a like-to-like showdown.

Initial Cost

Particulars	Nexon EV	Nexon Petrol
Ex-showroom price	INR 16,49,000	INR 10,94,900
Registration charges	INR 4,230	INR 1,20,440
Insurance premium	INR 70,933	INR 42,279
Other Charges, including TCS	INR 16,990	INR 12,949
Total initial cost	INR 17,41,153	INR 12,70,568

As you can note from above, although the registration charges are very low for EVs because of government incentives, the advantage is less due to the higher insurance premium on EVs. There is an expectation that insurance premiums could come down on EVs as they grow in popularity. There is an Rs, and 1.5 lakh income tax deduction on loan payments for EVs, so that should close the gap a bit more.

Regarding the upfront price, Nexon EV is costlier by Rs. 4,70,585.

Do the savings in running and maintenance costs in the EV version cover this price difference? Let's find out.

Assume that you clock in 15,000 km per year. So in 5 years, your car will cover 75,000 km.

Running cost for the EV version

The charge of electricity per unit in Delhi is Rs. 8 per kWh.(This varies from state to state, with many states charging lower than this). The battery of the Nexon EV has a capacity of 30.2 kWh, and therefore, a full charge will cost Rs. 242.

The range of Nexon EV on a full charge is rated 312 km under test conditions. Assume that in real-life situations, the same is 250 km. So the per km cost will be Rs. 0.97

For simplicity, if there is no increase in electricity cost over the years, the total running cost over five years or 75,000 km will be Rs. 72,750

Running cost for Petrol version

Nexon Petrol has a 44-liter fuel tank. The petrol price in Delhi as of August 2022 is Rs. 97 per liter. Therefore, a full tank will set you off by Rs. 4,268.

The fuel consumption of Nexon Petrol under test conditions is 16 km per liter. In real-life situations, the fuel consumption is 12 km per liter. The car will run 528 km on a full tank. So the per km cost will be Rs. 8.08

Assume there is no change in petrol costs over the years. Then the total running cost over five years / 75,000 km will be Rs. 6,06,000

Whoa! So the net savings on running costs over five years is Rs. 5,33,250. Add to that the lesser maintenance costs of EVs, which according to Tata Motors, is a saving of Rs 16,768, and the total savings on EV running cost is Rs. 5,50,018. This saving is well above the extra price of Rs. 4,70,585 that you paid for the EV version

You will save Rs. 79,433 over five years.

Although there are significant savings in running costs of EVs, the high initial cost is still a dampener. The higher initial price is mainly due to the price of the battery, although battery prices have fallen by more than 80% in the last ten years.

Let's say the battery prices go down further in the coming years, and the price difference reduces to Rs. 3,00,000 – which version of Nexon would you buy? For me, it's a no-brainer –

I'll go for the EV version.

Challenges for EV cost of ownership

Two critical challenges for reducing the EV cost of ownership are battery cost and battery life.

Battery cost

As pointed out many times earlier, the stumbling block of EVs remains the high cost of the battery pack. Though Lithium-ion packs for electric cars have fallen in price by more than 80% since 2010, they remain expensive.

There is a forecast for EVs to cost the same or less than a comparable gasoline-powered vehicle when the price of battery packs falls to about $100 kWh. Analysts say this price parity is achievable by 2024, while other studies have forecast the price of a lithium-ion battery pack will drop to as little as $80/ kWh by 2030.

Costs could come down drastically just with the scale of manufacturing. The next step would be to replace the graphite anodes in today's batteries with electrodes made of silicon. Silicon can store much more charge than graphite but expands and shrinks when charging and discharging and can react with and use up the electrolyte. These problems cause experimental silicon anode batteries to fail much sooner than those made with graphite.

Solving these issues could take five years, but it should boost energy densities to 350 Wh per kg.

The ICE cars will continue to see an upward price trend thanks to stricter emission norms and, therefore, installation of more expensive systems, and this should bring down the time taken for EVs to reach price parity.

Battery Life

There's also an issue with the battery life. Like the AA batteries we put into the TV remote control, Li-ion batteries eventually die. Even if you aren't using them, they'll begin to degrade. You can recharge them, but only a limited amount of times. It's like trying to fill up a pitcher of water with a tiny hole that grows bigger and bigger with each use.

We measure battery longevity in cycle lives or the number of times one can run it down, charge it up and use it again. With Li-ion batteries, starting from a 100% fully recharged battery will give you a longer individual cycle life but will reduce the total number of cycles you'll get. For that reason, the Tesla Roadster doesn't allow you to recharge more than 95 percent of the original power or let it drain down to less than 2%.

Currently, most manufacturers are offering 8-year/100,000-mile warranties for their batteries. This standard warranty is excellent, but remember to look at the fine print. Some manufacturers only cover the battery if it completely dies and cannot hold a charge, which does not happen often. Brands like BMW, Chevrolet, Tesla, Volkswagen, and Nissan will cover a battery pack if its capacity drops to a certain percentage, usually 60 to 70%.

While an electric vehicle's battery pack will tend to degrade slightly with each charge and discharge cycle, it's an extremely gradual process. For example, according to data compiled by the organization Plug In America, the battery pack in a Tesla Model S will only lose around five percent of its original capacity over the first 50,000 miles, with the rate of depletion slowing down from there.

The bottom line is that if it's properly cared for, an electric car's battery pack should last more than 100,000 miles before its range becomes restricted. Per the independent non-profit firm Consumer Reports, the average EV battery pack's lifespan is around 200,000 miles, nearly 17 years of use if driven 12,000 miles annually.

Let's look at the replacement cost of the battery pack of an electric car.

Tesla does not provide any official guide on the replacement cost of its battery. On the Model 3, with battery sizes ranging from 50 to 82 kWh, current battery price estimates put it between $7,000 to $11,000. The Model S's battery replacement cost is around $13,000-$14,000.

The cost of replacement of the 40 kWh battery pack in the Nissan Leaf will be around $6,500-$7500. The Hyundai Kona battery will also cost in the same range.

There are no verified battery replacement costs for the Nexon EV and the MG ZS EV in India, but there are reports that one would have to shell out anything between Rs. 6 to 8 lakhs for the battery packs. That is if it is out of warranty. This replacement is undoubtedly costly, but prices will decrease in time.

As long as you drive within 160,000 km during the 8-year warranty period, you don't have to worry about the battery replacement cost. And in 8 years, the battery costs are expected to come down substantially, so there's no point in fretting about the high prices.

Go on, enjoy the drive!

7

IMPACT OF ELECTRIC VEHICLES ON THE ENVIRONMENT

"You cannot escape the responsibility of tomorrow by evading it today."

—Abraham Lincoln

Fundamentally, two big reasons many countries are actively promoting electric vehicles are reducing pollution and dependence on oil.

The scourge of air pollution

The alarm bells over the dangerously high air pollution levels in major Indian cities are ringing so often that they have become numbing. Air pollutants, including particulate matter (PM), sulphur dioxide (SO2), nitrogen oxides (NO_x), carbon monoxide (CO), and ozone (O_3), often exceed the National Ambient Air Quality Standards (NAAQS) in Indian cities. According to a report called the Air Quality and Health in Cities, released by US-based Health Effects Institute in August 2022, India's capital New Delhi is the most polluted city in the world, with West Bengal's Kolkata bagging the second spot. Mumbai came in at the No. 14 spot.

The exhaust emissions from a running vehicle emit toxic substances such as Carbon Monoxide (CO), Hydrocarbons (HC), Oxides of Nitrogen (NO_x), Carbon Dioxide (CO_2), and Particulate Matter (PM). These deadly toxins can cause various health hazards, including congenital disabilities and adverse effects on the kidneys, lungs, and nervous system. PM10 emission from running exhaust and non-exhaust emissions (from the tire, brake, clutch, and road surface wear) is a major cause of breathing and respiratory diseases, lung tissue-related issues, cancer, and premature death.

One cannot ignore the problem of vehicular emissions, and the electrification of vehicles is one major initiative that the government of India is supporting.

Research has shown that electric cars are better for the environment. They emit fewer greenhouse gases and air pollutants than petrol or diesel cars. And this takes into account their production and electricity generation to keep them running.

The significant benefit of electric cars is their contribution to improving air quality in towns and cities. Pure electric vehicles produce no carbon dioxide emissions when driving with no tailpipe, reducing air pollution considerably.

Electric cars give us cleaner streets, making our towns and cities better places for pedestrians and cyclists. Over a year, just one electric vehicle on the roads can save an average 1.5 million grams of CO2!

Making electric cars does use a lot of energy. Even after considering battery manufacture, electric cars are still a greener option. The demand for an electric vehicle is because of the reduction in emissions created over the car's lifetime

The emissions created during the production of an electric car tend to be higher than a conventional car. This high emission is due to the manufacture of lithium-ion batteries, which are an essential part of an electric car. More than a third of the lifetime CO2 emissions from an electric car come from the energy used to make the car itself. As technology advances, this is changing for the better.

Reusing and recycling batteries is also a growing market. Research into using second-hand batteries is looking at ways to reuse batteries in new technologies such as electricity storage. One day we could all have these batteries in our homes to store our energy. Opportunities like this will reduce the lifetime environmental impact of battery manufacture.

Research by the European Energy Agencyfound that, even with electricity generation, an electric car's carbon emissions are around 17 – 30% lower than driving a petrol or diesel car. The emissions from electricity generation are also dramatically improved with the usage of low-carbon electricity.

Misconceptions about EVs

There are many myths surrounding EVs and their environmental impact. Let's look at some of the most common ones :

Myth #1: Electric vehicles are worse for the climate than gasoline cars because of the power plant emissions.

FACT: Electric vehicles typically have a smaller carbon footprint than gasoline cars, even when accounting for the electricity used for charging.

Electric vehicles (EVs) have no tailpipe emissions. However, generating the electricity used to charge EVs may create carbon pollution. The amount varies widely based on how local power gets generated, e.g., using coal or natural gas, which emits carbon pollution, versus renewable resources like wind or solar, which do not. Even accounting for these electricity emissions, research shows that an EV is typically responsible for lower levels of greenhouse gases (GHGs) than an average new gasoline car. To the extent that more renewable energy sources like wind and solar generate electricity, the total GHGs associated with EVs could be even lower.

Myth #2: Electric vehicles are worse for the climate than gasoline cars because of battery manufacturing.

FACT: The greenhouse gas emissions associated with an electric vehicle over its lifetime are typically lower than those from an average gasoline-powered car, even when accounting for manufacturing.

Some studies have shown that making a typical electric vehicle (EV) can create more carbon pollution than driving a gasoline car. This carbon pollution happens because of the additional energy required to manufacture an EV's battery. Still, over the vehicle's lifetime, total greenhouse gas (GHG) emissions associated with manufacturing, charging, and driving an EV are typically lower than the total GHGs associated with a gasoline car. That's because EVs have zero tailpipe emissions and are usually responsible for fewer GHGs during operation.

For example, Argonne National Laboratory researchers estimated emissions for a gasoline car and an EV with a 300-mile electric range. In their estimates, while GHG emissions from EV manufacturing and end-of-life are higher, total GHGs for the EV are still lower than those for the gasoline car.

Myth #3: There is not enough lithium and other rare earth material to meet the demand for battery production

A report confirms that the global supply of lithium-ion batteries is struggling to keep up with the increased demand from EV manufacturers.

However, this is a short-term problem. Perhaps reflecting an industry caught off-guard by the rapid increase in demand for EVs and batteries. Globally, battery production and the production of lithium and other precious metals are drying up rapidly.

Global lithium reserves are at over 14 million tons, and (depending on who you ask) the amount of lithium needed to meet current goals is between 0.5 and 1.3 million tons. In 2021, lithium extraction peaked at an industry record of 100,000 metric tons. These statistics can inspire confidence in the lithium industry's future, but there are many other reasons to feel bullish. Technological breakthroughs in the sector mean that lithium can be extracted more sustainably from brines in addition to increasing yields. In contrast, developing a lithium recycling industry is a foregone conclusion, ensuring that the world will not be in a position to run out of lithium.

Many studies set out to discover whether or not EVs are more harmful to the environment than ICE cars. Carbon Brief, for instance, compared the impact of driving 150 000 km in Europe with an ICE car vs. the 2019 Nissan Leaf. The results speak for themselves; EVs (the equivalent of 128g of C02 emission per km) pollute twice as less as the average ICE car (258g of C02 emission per km).

We reach a lower impact when considering more sustainable ways to produce batteries. For instance, a Tesla Model 3 running with a US-made battery will *only* account for an equivalent of 108g of CO2 emission per km. That's 58% less than a conventional ICE car.

Air pollution is a problem worth worrying about, but noise pollution can have a surprisingly adverse effect on human life. Not only does an excess of noise make our cities unpleasant to live in, but consistent loud noise also links to heart, sleep, or cognitive problems.

Transportation is mainly responsible for noise pollution, and the vast majority is due to road vehicles. This noise causes an innate stress reaction in humans, and we all know how terrible stress can be for us.

Electric vehicles make some noise when wheels interact with pavement, but they're relatively silent compared to conventional cars.

Considerations for Environment Impact

Four factors need consideration when evaluating the impact of electric cars on the environment. They are tailpipe emissions, well-to-wheel emissions, the energy source that charges the battery, and the car's efficiency.

When an electric vehicle runs on electricity, it emits no tailpipe (also known as direct) emissions. When evaluated on that factor alone, EVs are much more eco-friendly than conventional gasoline-powered vehicles today.

However, when evaluating the eco-friendliness of an electric vehicle, you also need to consider the 'well-to-wheel' (WTW) emissions. Well-to-wheel is an overarching term that includes greenhouse gas and air pollutants emitted to produce and distribute the energy used to power the car. Electricity production results in a varying amount of emissions depending on the resource. While 'being green' in driving your electric vehicle is a start, if your primary goal in purchasing an electric car is to reduce your greenhouse gas and pollutant emissions, you should also prioritize using zero-emissions electricity wherever possible.

When considering well-to-wheel emissions, all-electric vehicles emit around 4,800 pounds of CO2 yearly (US Department of Energy). Conventional gasoline cars will emit over twice as much annually – about 11,400 pounds of CO2 equivalent. So, if you hear anyone criticizing that EVs shift the carbon footprint from one point to another, you know it's a fib. The truth is that total electric vehicles generate half the average comparable gasoline car emissions.

The amount of well-to-wheel emissions your EV is responsible for largely depends on your geographic area and the energy sources most commonly used for electricity. Not all electricity sources are created equal; for example, the emissions would be higher when the electricity comes from coal or natural gas. The higher emission doesn't hold if your electric vehicle uses a primary electricity source like nuclear or solar power.

Outside of the resource used to produce your power, another reason why electric vehicles are considered more sustainable than traditional vehicles is the efficiency of the electric car. The gasoline in a conventional vehicle combusts to power the car, and 17 to 21 percent of the car's energy converts into fuel. On the other hand, they can convert 59 to 80 percent of the electric energy to power the vehicle.

According to Transport & Environment Organization, Europe's leading clean transport campaign group, all-electric vehicles manage an overall efficiency rating of 73%. It is 22% for hydrogen fuel cell vehicles and just 13% for standard fossil fuel vehicles using gasoline (See Figure on the next page).

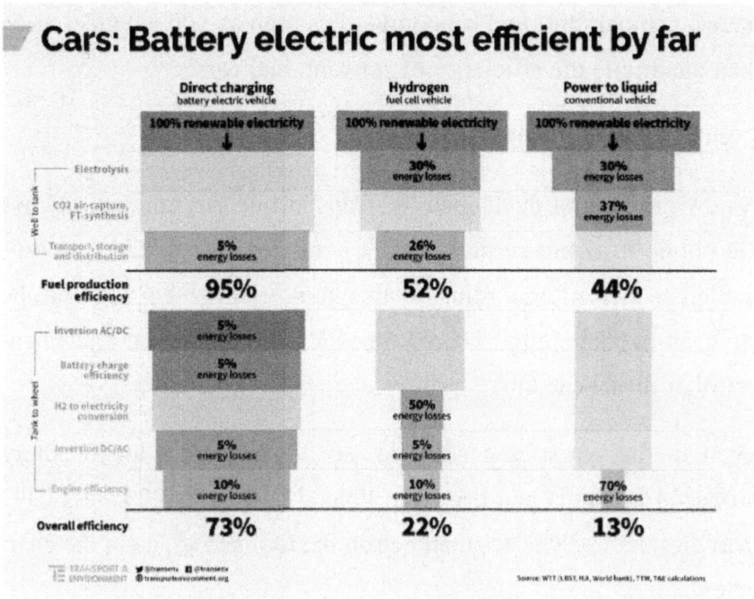

The study has concluded that adopting electric vehicles reduces air pollution for most emissions. The greater the percentage of renewable energy in the electricity mix, the more significant the reduction in air pollution.

When comparing an EV option to a conventional gasoline vehicle (or even a hybrid option), car shoppers also often evaluate MPGe, otherwise known as miles per gallon equivalent (of gasoline). The MPGe calculation represents the number of miles an electric vehicle can go given the same (or equivalent) amount of energy contained in one gallon of gasoline. The average MPG of a typical gasoline-powered car is around 24.8 miles per gallon. While that's much more efficient than in the past, it's not much compared to the MPGe of electric vehicles today. Electric cars available now can have a comparable 'fuel economy' of as high as 100 MPGe – more than quadruple the efficiency of conventional cars.

Comparison of lifecycle emissions

A new online tool developed by Transport & Environment allows the public to compare the lifecycle emissions of an EV to fossil-fuelled vehicles. According to this tool, electric cars in Europe emit, on average, almost three times less CO2 than an equivalent petrol or diesel car.

Even in the worst-case scenario, an electric car with a battery produced in China and driven in Poland still emits 22% less CO2 than diesel and 28% less than petrol, the tool shows, as in the chart below.

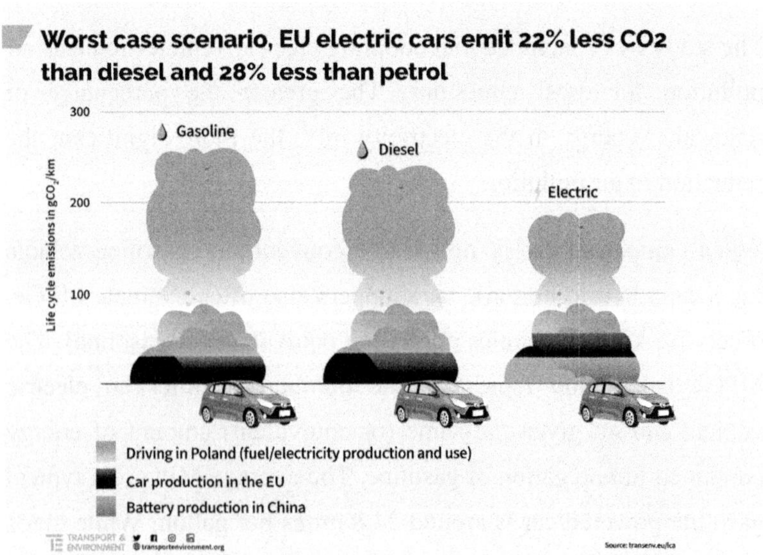

In the best-case scenario, an electric car with a battery produced in Sweden and driven in Sweden can emit 80% less CO2 than diesel and 81% less than petrol.

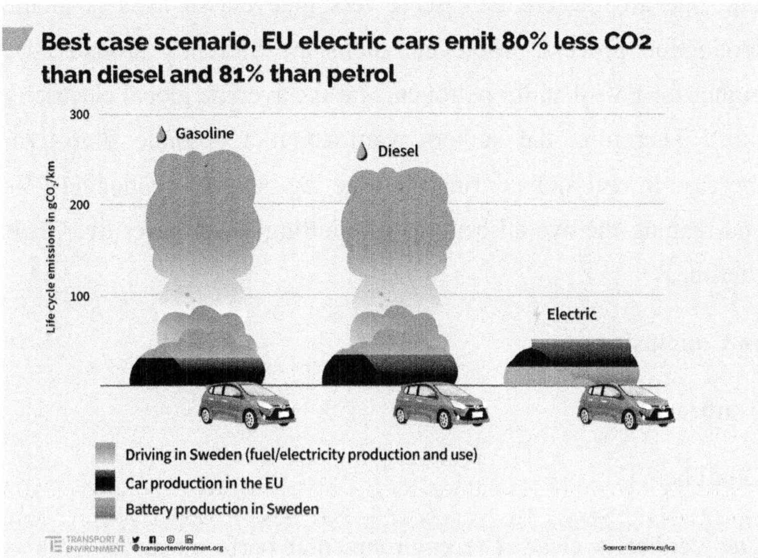

An independentstudy published in the *Nature Sustainability* journal by academics from the universities of Nijmegen, Exeter, and Cambridge has analyzed current and future emissions trade-offs involving electric cars and household heat pumps in 59 world regions. The authors find that 'current and future life cycle emissions from EVs and heat pumps are on average lower than those of new petrol cars and fossil boilers—not just on the global aggregate but also in most individual countries (53 out of 59).

The report says: Decarbonization would probably reduce global emissions even if end-use electrification did not get support from the rapid power sector. Sweden is a country using electricity mostly from renewable sources, making CO2 savings of around 70% on e-vehicles over petrol and diesel.

The one area where EVs score less than classic cars is in the production process, where emissions are currently around 30% higher for EVs than for petrol cars (at the average global electricity mix). Therefore, the authors warn about a possible short-term increase in emissions during the time lag between producing EVs and reaping the overall benefits of emitting fewer gases over their lifetime.

In Conclusion

Confused?

Don’t be.

The verdict is clear. The environmental performance of EVs is already better today than those of conventionally fuelled vehicles. The life cycle analysis carried out by Europe’s Transport & Environment, a private organization, shows that even when powered by the most GHG-intensive electricity in Europe, the carbon footprint of EVs is lower. As more renewable electricity enters the grid, the climate impact of EVs will further diminish. Likewise, technological improvement in battery chemistry, reusing batteries for storage purposes, and developing a recycling industry for EV batteries will improve their sustainability.

8

IMPACT ON OIL DEMAND

"The Stone Age didn't end for the lack of stone, and the Oil Age will end long before the world runs out of oil."

—Sheikh Ahmed Yaki Zamani

(Former Oil Minister, Saudi Arabia)

As electric car and truck sales continue to rise, demand for conventional internal combustion engine (ICE) vehicles will drop. Around 44.5% of global oil production makes gasoline or diesel for road transport, and the consequences for the oil industry are apparent. At the same time, electricity demand will increase significantly; for EVs to fulfill their central use case of being better for the environment, electricity will need to come from renewables.

Electric vehicles are starting to make a severe dent in oil use.

Scenario according to Bloomberg

In 2021, EVs displaced roughly 1.5 million barrels per day, according to a new analysis from Bloomberg New Energy Finance (BNEF). That is more than double the share of Evs in 2015. The majority of the displacement is coming from an unlikely source.

Two- and three-wheelers, not e-bikes, but mopeds, scooters, and motorcycles, have avoided oil use. Those EV options are especially prevalent in Asia, where they are rapidly adopted. According to Bloomberg NEF, electricity two- and three-wheeled vehicles — which tend to be popular in Asia—accounted for 67% of the oil demand avoided in 2021. Those vehicles had an outsized impact on oil demand. Next in rank were electric buses, which accounted for 16% of avoided oil demand, followed by passenger vehicles at 15%. The latter is on the verge of growing sharply.

By 2025, EVs of all types will displace almost 2.5 million barrels daily.

This news of displacement is undoubtedly a piece of good climate news. But it comes with a caveat: the amount of oil used displaced only a tiny share (3.3%) of the total global demand for transportation fuel, which landed at roughly 43.7 million barrels per day in 2021. Oil demand has to fall 37% by the end of this decade to keep the 1.5-degree-Celsius target in reach, according to the Intergovernmental Panel on Climate Change's 2018 report. Widespread EV adoption will play a huge role in determining whether doing so is possible.

According to BNEF, there are 12 million passenger EVs on the road today, but this will rise to over 600 million by 2040, based on markets, technological, legislative, and economic factors.

As a result of the increased adoption, Bloomberg estimates that EVs and fuel cells will displace over 21 million b/d of oil demand by 2050, with a net impact of 18.6 million b/d. This impact could still be higher if there are additional policies and regulations changes.

The changes in oil demand will vary from country to country, market to market, and even site to site within trade areas.

Bloomberg New Energy Finance estimates that road fuel oil demand will peak in 2027, but it will take another decade for the impact of advancements to be materially felt. Emissions will almost halve by 2050, but the sector will still be nowhere near net zero. In the best-case scenario, by the 2050s, fossil-derived road fuel demand will fall below levels last seen in the early 1970s. In this case, oil-related emissions will drop to 3.4 gigatons CO2 by 2050, down from almost 6.5Gt in 2019.

That said, the EV sector could end up hurting the oil sector in the long run, with BNEF predicting that electric and fuel cell vehicles will displace 21 million barrels per day in oil demand by 2050.

Passenger electric vehicle (EV) sales will overgrow in the next few years, rising from 6.6 million sold in 2021 to 21 million in 2025. The fleet of EVs on the road will hit 77 million by 2025 and 229 million by 2030, based on BNEF's Economic Transition Scenario. That's growth from 16 million at the end of 2021, reflective of the remarkable success story of EVs in the energy transition.

Forecast by other analysts

A report from IHS Markit shows that in 2020, light plug-in and fuel-cell vehicles, as well as electric city buses and two-wheelers, collectively displaced about 370,000 barrels per day of global oil consumption. This figure will grow to 1.5 million barrels per day by 2025, equal to about 1.4% of the projected level of total world oil demand.

According to McKinsey, global liquids demand will reach around 104 million b/d in the next two to five years, driven primarily by the uptake of electric vehicles. However, crude oil demand will decline rapidly only beyond 2030, and 70% is the expected decline in liquid demand in road transport by 2050, after peaking in the early 2020s.

The International Energy Agency (IEA), which represents 29 oil-importing industrial countries, produces bellwether forecasts that foresee electric cars phasing in slowly. Its baseline projection envisions 150 million electric vehicles on the world's roads by 2030, or about 10 percent of all passenger vehicles at that point. In comparison, only two million electric vehicles are operating today – 0.2 percent of the 1.2 billion on the road. The IEA estimates this shift will save nearly two million barrels of oil per day, relative to its business-as-usual projection of the world using at least 70 million barrels per day for transportation by 2040. That consumption level would mark a 30 percent increase from roughly 54 million barrels.

Given the dominance of internal combustion engine passenger vehicles, including cars, SUVs and light trucks, replacing them with electric models will take decades. Automobiles are durable goods that typically remain on the road for 10 to 15 years. Not all drivers will soon buy a new car, let alone an electric one.

In other words, even if (hypothetically) all new car sales were to turn electric instantly, it would likely be sometime after 2030 before gasoline cars would disappear. Besides, passenger vehicles consume only about 26 percent of the oil used worldwide. Given these stubborn realities, these electric vehicles still represent a tiny portion of new-car sales; reaching a peak in oil demand by 2040 would require more than widespread conversion to electric-powered cars.

But with other trends taking shape, electric vehicle growth could potentially revolutionize transportation enough for oil consumption to stop growing within this time frame.

Even if all of Europe mandated that only plug-in vehicles would be on sale starting in 2030, and China followed suit by 2035, that wouldn't bring about peak oil demand by 2040. According to IEA's research, global oil consumption will keep growing until as late as 2050, partly because so many cars and trucks running on gasoline and diesel – especially in developing countries – will remain in use.

Predictions by BP Global

BP Global, the British Energy giant, has put forth three scenarios in their Energy Outlook 2022 to explore the energy transition by 2050. Accelerated, Net Zero, and Momentum. These include various possible levels of fall in CO2 emissions in the future to paint a range of scenarios. While New Momentum assumes the broad trajectory along which the global energy system is progressing, Accelerated and Net Zero provide a more substantial reduction in carbon emissions.

BP's forecast projects a fall in oil demand in all three scenarios, driven by the declining use of road transportation. Even in the New Momentum scenario, the oil demand will fall to 80 Mb/day in 2050 from the current level of about 100 Mb/day. In the Net Zero scenario, the oil demand will fall drastically to just about 25 Mb/day in 2050.

The decline in oil consumption is largely due to the falling use of oil in road transport as the fleet becomes increasingly electrified and more fuel-efficient. In all three scenarios, the decline in oil demand accelerates throughout the outlook as the electrification of road vehicles progresses.

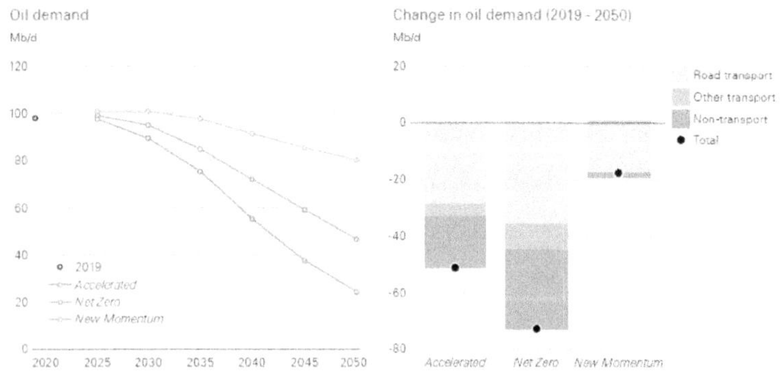

The dominant factor in reducing oil consumption in road transport over the first part of the outlook is the continuing improvement in vehicle efficiency, primarily passenger cars. By 2035, progress in vehicle efficiency will account for 60-70% of the fall in oil used in road transport across the three scenarios.

The other major factor reducing the oil in transport is the electrification of the vehicle part. Electric Vehicles in Accelerated and Net Zero account for 65-80% of the vehicle kilometers traveled on the road in 2050, compared with less than 1% in 2020.

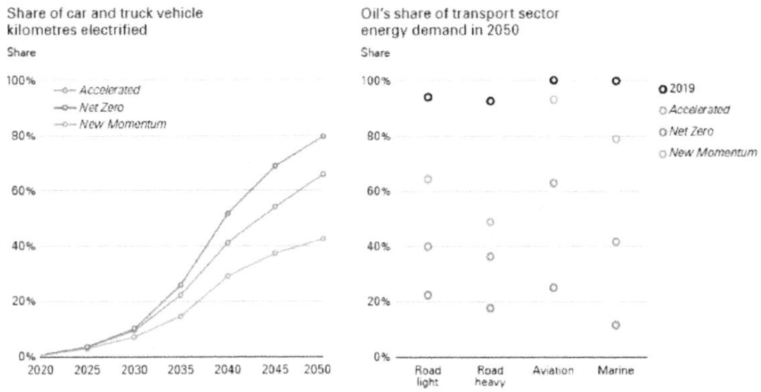

Demand forecast by OPEC

In its 15th Edition of World Oil Outlook 2021, OPEC has projected a much rosier picture than the doom and gloom painted by BP Global. As per their report, oil demand will increase by 17.6 mb/d between 2020 and 2045, rising from 90.6 mb/d in 2020 to 108.2 mb/d.

The long-term projections highlight a contrasting picture between ongoing demand growth in the non OECD region and declining demand in the OECD. The need for oil in OECD countries will fall from 42.1 mb/d in 2020 to 34.2 mb/d in 2045. At the same time, the oil demand in non-OECD will rise to 74.1 mb/d in 2045 from 48.6 mb/d in 2020.

The main reasons for the declining trend in OECD countries are efficiency and improvements across all consumption sectors and the substitution of oil for gas and renewable energy. This decline includes the significant penetration of EVs in the road transportation sector, ongoing electrification of residential and industrial sectors, and the penetration of alternative fuels in the marine and aviation sectors, among others. This picture includes a static (and aging) population and low economic growth, with less oil-intensive industry, especially in the second part of the forecast period.

In contrast, demand will continue to grow in the non-OECD region, driven by an expanding middle class, high population growth rates, and more substantial economic growth potential.

OPEC also forecasts that the oil demand by the road sector will rise from 44.6 mb/d in 2020 to 46.2 mb/d in 2045. One big reason is that the global passenger car stock will try to surpass 2 billion units by 2045, almost double that of 2020, which was about 1.2 billion. Most of this increase will come from non-OECD countries, which will see their passenger vehicle stock grow exponentially from 551 million units in 2020 to nearly 1.4 billion units in 2045.

The global commercial vehicle stock will also double from 248 million units in 2020 to 503 million units in 2045.

According to OPEC's WOO 2021, electrification of the global fleet will grow steeply in the medium and the long term. It will attain about 10% in around 2035 and rise to almost 20% by 2045. It is an assumption that by 2045, the on-the-road EV fleet will approach 500 million vehicles.

But again, there will be opposing trends in oil demand by the road transportation sector between OECD and non-OECD countries. The oil demand by the road transport sector in the OECD region will fall from 23.5 mb/d to 14.3 mb/d because of the shift to electric vehicles. On the other hand, this decline will be more than offset by an increase in oil demand of the road transport sector in the non-OECD region from 21.1 mb/d in 2020 to 31.9 mb/d in 2045. That is not to say that EVs won't impact the on-OECD area. The fleet expansion would add much more oil demand *ceteris paribus* but will reduce by penetration of EVs and efficiency gains.

Tony Seba's Death Spiral

Futurists make predictions. That's what they do. Sometimes they are right; sometimes, they are wrong. As futurists go, Stanford economist and RethinkX founder Tony Seba have a pretty good track record. He predicted the solar energy boom when prices for solar power were ten times what they are today. His latest report predicts two things. He says that by 2030, 95% of people won't own a private car, killing off the auto industry. Two, he predicts electric vehicles will devastate the global oil industry by the same date.

Seba's thesis is dependent on the mass adoption of TaaS - Transportation as a Service - with most people giving up driving and car ownership.

Tony Seba insists, "On-demand electric autonomous vehicles will become ten times cheaper than owning a vehicle. This demand will make economic sense across the globe whether it's India, Africa, the US, or Europe, and this will be a global disruption."

Seba says the impact on the oil industry will be "catastrophic." He predicts, "Global oil demand will peak at 100 million barrels per day by 2020, dropping to 70 million barrels per day by 2030. This decline will disproportionately impact different companies and countries and, in many cases, dramatically depending on their exposure to high-cost oil. Essentially, that means that the new equilibrium price in the oil markets will be $25. So, if you produce oil and can't compete at $25, you are essentially holding stranded assets."

Seba's predictions are by far the most dramatic and outrageous, leaving many automotive old guards and oil giants in total disbelief. It makes for a wild ride ending with a new paradigm almost unrecognizable from today.

The revolution brings significant risks for incumbents with hefty sunk costs and high expectations for return on investment. "Disruption usually happens from the outside. Kodak did not disrupt itself; it was other companies that did it. Newspapers did not disrupt themselves; it was web companies – the Googles and the Facebooks," noted Seba.

9

EFFECT ON AUTOMOBILE DEALERSHIPS

"If you always do what you've always done, you'll always get what you've always got."

—Henry Ford

Dealerships profit much from post-sales maintenance, and electric cars require much less post-sales maintenance. Dealership business models don't add up for electric vehicles.

The growth of electric mobility will have far-reaching consequences for auto dealers, touching virtually every aspect of the business. There's much to consider, from redistributing profit pools to adopting new operating procedures. Maintenance workers will need the training to service high-voltage powertrain systems safely and efficiently. At the same time, frontline sales staff will competently answer consumer questions and close the sale. Electric vehicles (EVs) also tend to have fewer mechanical parts that break down, which has significant implications for after sales service.

It seems clear that auto dealers will sell more EVs in the coming years, significantly affecting their profits from both front-end business and back-end operations.

Impact on front-end sales

Today, a car dealership struggles between the higher costs of EV manufacturing and the need to keep vehicles priced competitively. Consequently, average dealership sales margins for EVs are lower than for ICE vehicles. Moreover, a dealership's power to influence sales margins may face problems from an OEM's increasing power in pricing controls and online channels.

As more traditional OEMs commit to increasing EV production, the desire to improve online sales channels or even to sell exclusively online has intensified. Some direct-to-consumer models still include the dealership (for example, purchase online, pick up at the dealership).

Impact on back-end service and parts

Many dealerships rely on finance reserves, or the portion of the buyer's interest rate earned by the dealership for securing the car loan, as a source of revenue. EVs are typical \$5,000 to \$10,000 more expensive than comparable ICE vehicles. As a result, more buyers will likely need the help of the dealer's finance department, which should result in more financing deals and lease volume for EVs. Furthermore, there is an expectation that the share of lease deals will increase until confidence in batteries goes up. In addition, for savvy dealers that partner with local installation services, there might be opportunities to offer financing contracts for car batteries and at-home charging equipment to consumers.

With more financing and leasing deals for EVs, the back office may also find that consumers are more willing to add on various F&I products, such as extended service contracts, since many early customers will likely be from more affluent segments. Guaranteed asset-protection insurance, which covers any difference between the value of the vehicle and its special financing in the event of a total write-off, will be most appealing to consumers, along with personalized electronics and battery-maintenance plans.

Auto dealers will likely experience a decrease in after sales parts revenue since EVs have fewer mechanical parts that break down. According to one study by McKinsey, EVs will probably generate up to 40 percent less aftermarket spending when compared with similarly aged ICE vehicles.

The most significant differences between EV and ICE aftermarket spending will likely emerge in general repairs, such as brake pads that tend to wear and tear and consumables (for instance, fluids and traditional powertrain parts). However, due to the technical layout of EVs, other chassis components, EV powertrain parts, thermal parts, and tires will see a surge in spending in the long term.

The impact of electrification on service revenue is less specific. Compared with ICE vehicles, EVs experience lower wear and tear per mile traveled, typically resulting in lower maintenance costs.

However, even though EVs typically require less maintenance work than ICE vehicles, with less frequent service touch points, servicing EVs requires specialized capabilities, as the tasks involved are more complex. Because of this, each maintenance visit may necessitate more billable service hours and generate increased revenue per hour. And dealers might offer new service offerings. For example, increasing penetration of advanced driver-assistance systems (ADAS) means that more vehicles may need to undergo a typically lengthy and complex sensor-calibration process. Of course, getting the service business ready for EVs will also require investment.

Finally, over-the-air (OTA) update capabilities could further affect dealers' after sales revenues by reducing service traffic to the dealership. Most current service work, such as brake service, tire rotation, and suspension repair, require an in-person visit to the dealership. However, new cars—especially EVs—contain an increasing share of software and electronic components. The electronic components will enable carmakers to offer more OTA services, such as wireless updates to the navigation and entertainment systems and software upgrades, removing the need for new car owners to visit the dealership in person for those improvements.

In summary, EVs could considerably strain dealers' aftermarket sales and profits. EVs contain fewer moving parts and require less maintenance work, which could significantly reduce after sales parts revenue for EVs compared to ICE vehicles over the next decade. At the same time, dealerships will need to meaningfully invest in preparing maintenance staff to work on EV powertrains, including specialized training and purchasing high-voltage tools and safety equipment.

How auto dealers can adapt

Changing market dynamics and EV sales growth mean dealerships would be wise to act quickly to address all elements of the EV purchasing journey. In doing so, they can capture new opportunities.

Train the frontline and transform the sales journey. Dealers can continue digitizing the sales journey and offer a seamless, omnichannel experience that touches on the EV itself and the broader ownership experience, such as charging and service. They can enhance the customer experience by including EV-specific elements within the showroom (for example, the fast-charging expertise). They can also provide an end-to-end offering for customers that makes EV ownership as simple as possible (for instance, selling preferred installation services for at-home charging equipment).

Dealers can also continue to train sales associates or designate EV experts to learn about buying criteria such as range, vehicle performance, service expectations, and day-to-day operation. They might fine-tune each step of the decision journey to cover all aspects of EV ownership (for example, adapt the test drive to include charging the vehicle and vehicle performance). As the EV market expands, an adequately trained frontline sales staff can skillfully guide consumers to the right vehicle.

To capture growing customer demand, dealers could adjust vehicle inventory to include more new and used electric vehicles. They might employ data and analytics to optimize inventory further to achieve the most competitive stock at the trim level. For example, by tracking broader marketing inventory and trends, dealers can optimize their inventory to focus on high-demand and high-margin vehicles and trim levels. In addition, by offering comprehensive subscription models as an affordable entry into EVs—for instance, an extended test-drive phase or a short-term ownership solution such as a month-by-month lease—dealerships could reduce cost barriers to EV ownership.

Reshape F&I. Auto dealers could investigate new Finance & Insurance products tailored to EV-specific maintenance needs. The extended warranties on the battery, emergency charging services included in roadside assistance, or tires-as-a-service with a monthly subscription makes the more significant expense easier to pay to lower the barrier to ownership and TCO. Moreover, dealerships should consider rethinking the residual values of ICE and BEV vehicles. As BEV adoption continues to grow and demand increases over time, low resale values could potentially erode the attractiveness of ICE vehicles in the future. We see additional opportunities to offer an end-to-end EV leasing or financing solution that includes everything from financing the car, home charger installation with utility partnerships, and seamless commercial charging access, all in one monthly payment.

Reform your service concept. For service and reconditioning centers to effectively service a growing number of EVs, auto dealers will need to establish new point-of-sale processes. They need to train technicians and other service personnel to manage high-voltage systems safely and invest in the special tools and equipment required to service electric powertrains. Consequently, auto dealerships will need increased investments in EV after sales readiness, given the new tools, capabilities, and learning curve. Some OEMs have already adjusted their dealer contracts to reflect the shift toward electric-vehicle sales by setting dedicated targets requesting defined charging infrastructure and stipulating different processes such as battery maintenance and handling.

In addition, auto dealers might assist new, disruptive OEMs and digital dealerships. This assistance would increase the utilization of assets for traditional dealerships while also lowering the barrier to entry for the new EV-focused OEMs and digital disruptor retailers that don't own service networks. Dealers could also expand services not affected by electrification, such as tire sales and seasonal tire storage, and rethink how these services can be given - for instance, through OTA status updates or at-home services - to continue servicing through the dealership attractive for customers.

Furthermore, dealers could consider partnering with manufacturers to invest in or subsidize EV-specific infrastructure or hardware, such as charging infrastructure for customers and the service shop, new tools and parts for service, or project coordination. Additionally, dealers might establish partnerships with local retail chains or energy providers to offer charging-infrastructure installation services for critical use cases, such as fleets owned by emergency responders. Dealerships could also pursue EV certification to drive more business to service bays.

Expand into the fleet. Dealerships might rethink how to resell used vehicles and direct them. For example, ride share workers or people interested in working in ride sharing. This ride sharing would enable dealerships to reduce aging inventory costs while attracting potential new buyers. Dealers might also offer fleet services for small- and medium-scale fleet owners, focusing on providing low-cost maintenance and repairs through trusted service networks.

Reassess your dealership strategy. Dealers could expand their used-car inventory to include off-brand EVs to generate more traffic and become a trusted EV store across multiple brands and models. In addition, dealers may wish to redesign the contract model between OEMs and dealerships, for example, by allowing online orders to be filled up by dealerships.

Redefining the operating model

Evidence from leading fleet companies suggests that due to their simplified mechanicals, lightly used braking systems, and higher levels of reliability, the service, maintenance, and repair (SMR) costs of EVs are around half those of petrol and diesel vehicles. In addition, most electric cars have longer service intervals, while some, such as Audis, have no mileage limitation during each two-year scheduled service period.

Several practical reasons to consider are: electric vehicles have fewer moving parts, brakes tend to be replaced less frequently due to regenerative braking, service intervals are usually longer, and fewer billable hours.

It is important to note that despite the diminishing returns, retailers still make significant investments in staff, EV training, tools, extra EV charge points, and EV courtesy vehicles.

One thing is clear; dealerships are operating within an increasingly disrupted environment, which has affected the bottom line and created some uncertainty for the future. Over the past several years (except the COVID-19 pandemic), dealerships have experienced margin compression on vehicle sales. With threats to their services business, margin compression could continue. Higher front-end margins, notably F&I, will come under further pressure as EV prices decline and battery costs decrease.

The good news is most EV OEMs require factory-authorized dealership service departments for repair and maintenance. Further, even though independent shops handle 70% of aftermarket service of ICE vehicles, franchise dealers don't want to cede EVs to them, especially as consumers familiarize themselves with battery charging and other peculiarities. The EV owner might trust the dealers more to perform service than the aftermarket shops earlier in their ownership period. The threat of DIY and independent service center can have limitations in the near term.

Many dealerships are embracing the EV transition. While there are fundamentals to guide dealerships over the coming years, there are many uncertainties and unanswered questions. Dealerships should develop a business strategy, shift their operating model, and make a road map to address these uncertainties.

Regarding strategy, the critical question centers on the degree of scale necessary to compete and grow profitably.

- In which portfolio of brands to invest?
- How many stores to develop and in which markets?
- Whether to acquire other dealerships?

In redefining the operating model, dealerships must focus on creating the best customer experience efficiently and effectively.

- How to enable this through the optimal digital and omnichannel strategy in collaboration with the OEMs?
- Should subscription services be offered for brand and vehicle portfolios and maintenance program bundles?
- To what degree do you invest in charging infrastructure to reduce electricity costs and create new revenue streams by selling electricity back to the grid or providing customer service? What role does solar play in this approach?
- How to fully utilize the federal, state, and local incentives?
- How to ensure the number of required certified technicians?

Once dealerships have answered these questions, prioritize the opportunities and organize them into a road map to guide the transition. For any investments required, a clear and tangible business case should properly filter out those initiatives if it has to be done or not.

With EVs requiring less routine maintenance, dealers can focus more on specialty services for their EV customers. This area offers the best revenue growth opportunity, considering that this market is currently small but growing.

Dealers can offer home charging equipment for EVs. No longer do buyers need to wait in long lines at gas stations. They can install chargers at their homes, opening up other opportunities for dealerships to provide services other than just selling the vehicles. They can also focus on post-sale services such as battery replacements and software updates. This area is critical as technology changes, and newer models emerge yearly.

Dealerships could also consider the services offered by EV manufacturers. Some companies provide pickup and delivery services, while others will even recharge vehicles at their facilities for a fee. This service is an excellent way to earn revenue without selling something that may not be necessary. As more EVs hit the road, dealerships will need to invest in long-term service and support. This long-term support includes warranty services for new EVs on top of routine maintenance.

Dealerships should prepare for this market's expansion by optimizing their IT infrastructure to handle such requirements adequately. It also means that dealerships need to keep up with EV technology. Manufacturers are continually improving their vehicles, which could also improve the battery charging rates. Thus, dealerships should be ready to invest in newer charging stations and other equipment.

Dealerships also need to think of ways to offer the best service to keep their current customers happy and loyal while attracting new ones. Besides providing more extended warranty periods for EVs, they can focus more on safety features such as emergency stop assistance and backup cameras. This way, car buyers know that their EVs will be safer on the road.

The message to car dealerships – innovate or die.

If projections for EV adoption are accurate, car dealerships clearly won't be able to remain who they are for much longer. When considered alongside other disruptions roiling the auto industry—car sharing, on-demand, connected vehicles, self-driving cars—electric vehicles may be the domino that accelerates the upending of traditional dealership networks.

There is time. The introduction of EVs into the mainstream is a long road, and dealerships can take steps to stay relevant when the deluge swamps the market. The auto technician of the future will need to become an IT support guy with a car lover's mind, someone with the ability to change tires and operate diagnostic and scanning equipment to root out problems involving computer networks and data processing.

There is a silver lining to the service transformation due to EVs. The sophisticated electronics in EVs will give authorized service centers an edge over roadside mechanics and unauthorized workshops. The current service retention for OEM dealerships in India is between 40-45 percent after the warranty period ends. As more EV customers visit the authorized service centers to get their batteries and electronics serviced, this will go up in the future, thus driving volumes for the dealers. Hence, OEMs and dealers need to develop a roadmap for re-skilling their workforce to capitalize on this future opportunity.

As has happened in many industries, which have been either wiped out or greatly diminished by technological changes, there would be a majority who will bury their heads in sand and hope for the storm to pass over. The smart ones will build a new shelter and no prize for guessing who will survive.

10

TESLA, INC

"When something is important, you do it even if the odds are not in your favor."
—Elon Musk

When you think of an electric car, you think of Tesla and Elon Musk. Tesla is iconic, and Musk is a legend.

Barely a decade old, Tesla is already gigantic and adored. Its market capitalization hovers above $900 billion as of August 2022.

When Tesla Model 3 opened for bookings in 2016, it received a mind boggling pre-order of 400,000! The Model 3 is the all-time best-selling plug-in electric car worldwide. In June 2021, Tesla became the first electric vehicle to sell 1 million units globally. Tesla's global sales were 936,222 cars in 2021, an 87% increase over the previous year, and cumulative sales totalled 2.3 million cars at the end of 2021. In October 2021, Tesla's market capitalization reached $1 trillion, the sixth company to do so in U.S. history.

Tesla has almost single-handedly brought electric vehicles into the mainstream.

Contrary to popular belief, Tesla is not an overnight success, and it is no longer a start-up as it is a pioneer in the electric car market. Many people think that Elon Musk is the founder of the company. But this is far from the truth. Two Silicon Valley engineers, Martin Eberhard and Marc Tarpenning, who, according to the company website, "wanted to prove that electric cars could be better than gasoline-powered cars," founded the company in 2003.

A Brief History

The company's seeds go back to 1990 when Marc Tarpenning met Martin Eberhard, then an engineer at Wyse Technology became good friends. Tesla was the brainchild of Eberhard. The two had much in common, including a passion for starting companies, and they soon launched companies, including NuvoMedia, a company developing eBook readers. A passion for autos started soon after Eberhard went through a divorce and decided to buy a sports car. He invested in a company, a boutique electric automaker called AC Propulsion, and wondered if this was a market he could explore and work.

In 2001, Eberhard and Tarpenning met Musk when they heard him speak at a Mars Society talk at Stanford University and introduced themselves. Musk came with a successful history of starting up companies, and he, along with Peter Thiel and Max Levchin, was the co-founder of PayPal. The takeover of Paypal by eBay happened in 2002. This takeover led to the creation of a new company called Space X, which designs, manufactures, and launches advanced rockets and spacecraft.

A few years later, their paths would cross again when Eberhard and Tarpenning asked to meet Musk to share their electric car idea with him. The three met to discuss the concept with Musk on board.

Tesla invented the electric car that was powerful and beautiful with zero emissions and. was officially incorporated in 2003. The company was named after electrical engineer Nikola Tesla and aimed to "accelerate the world's transition to electric mobility with a full range of increasingly affordable electric cars." In February 2004, via a $6.5 million investment, Elon Musk became the company's largest shareholder. Other co-founders were JB Straubel, who served as the CTO at Tesla until moving to an advisory role in July 2019, and Ian Wright, who left Tesla in 2004 and founded another company, Wrightspeed.

Between 2004 and 2008, Tesla continued to grow and develop its first automobile, the Roadster. The company opened its manufacturing plant in Fremont, California, a 5.5 million square foot factory that used to be owned by Toyota and General Motors. The factory is known as Nummi and includes two paint facilities and 1.5 miles of assembly lines.

Martin Eberhard was the original CEO of Tesla until the board of directors asked him to resign in August 2007. Eberhard then took the title of "President of Technology" before ultimately leaving the company in January 2008 along with co-founder Marc Tarpenning. The latter served as the company's Vice President of Electrical Engineering until then.

Musk became the company's CEO in 2008 and product architect, positions he still holds. That same year, Tesla launched its first car and sports car, the Roadster. "It is not just a car, but one of the strongest automotive statements on the road," Car and Driver wrote. The Executive summary reveals -the company has always focused on the car's mechanics as much as the design. "High performance," defined in the executive summary, included going from 0-60mph in less than 3.9 seconds and zero maintenance for up to 100,000 miles other than tires.

Then came the October 2008 financial crisis, and Tesla almost collapsed. It fired 25% of its staff. Tesla was on the brink of bankruptcy, and it began to struggle to raise $100 million to fund its Model S. At the time, Elon Musk put his last $35 million into the company.

The savior was the $8 billion Advanced Technology Vehicles Manufacturing Loan Program. Tesla secured about $465 million in a low-interest loan from the Department of Energy in 2009.

The expenditure is $365 for the Model S car and the remaining $100 million for a powertrain manufacturing plant. Tesla repaid the loan in full in May 2013, with $12 million in interest.

On June 29, 2010, Tesla launched its initial public offering (IPO) on NASDAQ. For US$17.00 per share,13.3 million shares of common stock got distributed among people. The IPO raised US$226 million.

The meteoric rise of Tesla

The Tesla Roadster (2008) was the first production automobile to use lithium-ion battery cells and the first production EV with a range greater than 200 miles (320 km) per charge. Between 2008 and March 2012, Tesla sold more than 2,250 Roadsters in 31 countries.

In May 2010, Tesla purchased what would later become the Tesla Factory in Fremont, California, from Toyota for $42 million and opened the facility in October 2010 to start production of the Model S.

In January 2012, Tesla ceased production of the Roadster, and in June, the company launched its second car, the Model S luxury sedan. The Model S won several automotive awards during 2012 and 2013, including the 2013 Motor Trend Car of the Year, and became the first electric car to top the monthly sales ranking of a country when it achieved first place in the Norwegian new car sales list in September 2013. The Model S was also the best-selling plug-in electric car worldwide in 2015 and 2016.

Tesla began shipping its third vehicle, the luxury SUV Tesla Model X, in September 2015, when it had 25,000 pre-orders.

Tesla began selling its fourth vehicle model, the Model 3 sedan, in July 2017. The Model 3 was cheaper than previous Tesla vehicles and intended to cover the mass market. It was highly anticipated, which prompted the company to try to speed up production. By August 2017, there were 455,000 reservations for the Model 3. Delays and production problems plagued the rollout. The rollout problem increased pressure on one of the market's most sort after companies.

By the end of 2018, there was no production problem, and the Model 3 was the world's bestselling plug-in electric car for the year.

In 2021, Tesla sold over 500,000 Model 3 sedans 2021, up 40 percent year over year! The Tesla Model 3 was the eighth best-selling car in the world for 2021. What a meteoric rise!

Tesla opened its first "Gigafactory" outside the United States in Shanghai, China, in 2019. Giga Shanghai was the first automobile factory in China wholly owned by a foreign company, and the factory construction just took six months to come into operation. The following year Tesla also started construction on a new Gigafactory near Berlin, Germany, and another in Texas, United States.

In March 2020, Tesla began delivering its fifth vehicle model, the Model Y crossover.

At Tesla's annual shareholder meeting in early August 2022, Musk announced that the Tesla Model Y is on track to become the world's best-selling vehicle. More specifically, the electric SUV will be the world's best-selling vehicle by revenue this year. The company expects it to be the best-selling vehicle by volume next year once Tesla has ramped up production at Gigafactory Texas and Gigafactory Berlin.

The best-selling vehicle in the world is Toyota Corolla, with around 1,150,000 sales in 2021.

Tesla doesn't break down sales between the Model 3 and Model Y, but expectations are that they have surpassed the former. Tesla is reporting a capacity of around 300,000 of those vehicles per quarter. At Gigafactory Texas and Berlin, Tesla is exclusively producing the Model Y. By the end of 2022, Tesla will have the capacity to produce over 1 million Model Y vehicles per year as per their expectation from the market.

Future Products

Roadster (2nd Generation): In a surprise reveal at the end of the event that introduced the Tesla Semi on November 16, 2017, Tesla unveiled the second generation Roadster. Musk said that the new model would have a range of 620 miles (1,000 km) with the 200 kWh battery pack and achieve 0–60 miles per hour (0–97 km/h) in 1.9 seconds. It also will reach 0–100 miles per hour (0–161 km/h) in 4.2 seconds, and the top speed will be over 250 miles per hour (400 km/h). The SpaceX Package will include cold air thrusters that will increase the pace even more. The vehicle will have three electric motors allowing for all-wheel drive and torque vectoring during cornering. Deliveries would start in 2023.

The Tesla Semi is an all-electric Class 8 semi-trailer truck announced in November 2017. Musk confirmed that two variants would be available: one with 300 miles (480 km) and one with 500 miles (800 km) of range. The Semi will get the power from four independent electric motors of the type used in the Tesla Model 3. It will include an extensive set of hardware sensors to enable it to stay in its lane, a safe distance away from other vehicles, and later, when software and regulatory conditions allow, provide self-driving operation on highways. The deliveries were to begin in 2019, but now they will be delivered in 2023, primarily because of battery supply issues.

Cybertruck: The Cybertruck is a pickup truck unveiled on November 21, 2019. On September 22, 2020, Musk revealed roughly 600,000 Cybertruck preorders. Production of the Cybertruck is delayed, and may begin only by end of 2023.

Tesla's strategy

Tesla aims to disrupt the automotive industry by creating many innovative pieces that fit together. This strategy was called "complex coordination" by Tesla investor Peter Thiel. Its marketing, production, sales, and technology strategies differ notably from its competitors.

Tesla's automotive strategy is to emulate typical technological product life cycles and initially target affluent buyers. It would then move into larger markets at lower price points. Each model's battery and electric drivetrain technology will develop and pay for through the earlier models' sales.

The Roadster was low-volume and priced at US$109,000. Model S and Model X targeted the broader luxury market, and model 3 aimed at a higher-volume segment. This business strategy is prevalent in the technology industry. According to a Musk blog post, "New technology in any field takes a few versions to optimize before reaching the mass market, and in this case, it is competing with 150 years and trillions of dollars spent on gasoline cars."

Tesla's production strategy includes a high degree of vertical integration (80% in 2016), which provides for component production and proprietary charging infrastructure. The company operates enormous factories to capture economies of scale. Tesla builds electric powertrain components for vehicles for other automakers, including the Smart ED2 Fortwo electric drive (the lowest-priced car from Daimler AG), the Toyota RAV4 EV, and Freightliner's Custom Chassis Electric Van. Vertical integration is rare in the automotive industry, where companies typically outsource 80% of components to suppliers and focus on engine manufacturing and final assembly.

Tesla's business model includes a three-pronged approach to selling, servicing, and charging its electric vehicles.

Tesla doesn't adopt the approach of franchise dealerships, unlike most manufacturers. They prefer selling their product directly to customers through self-owned showrooms across many of the major urban centers in the world. They believe that this method of selling can speed up their product development. But more significant is the customer's buying experience. Tesla has showrooms, Service Plus centers (retail and service centers), and service facilities. Tesla has also made use of Internet sales—consumers can customize and purchase a Tesla online.

Customers can service their cars or charge them at the service centers or the Service Plus locations. They also have mobile technicians called Tesla Rangers who can come to your home. With the Model S, they can wirelessly upload data, so technicians can view and fix specific problems online without even physically touching the car.

Tesla has a vast network of locations where its customers can charge their vehicles. Supercharger Stations provide a place where customers can charge their cars for about 30 minutes for free. They believe this will increase the customer's product adoption rate.

It doesn't end here. Besides the three-pronged business model, Tesla also provides financial services like granting loans and leases. If the customer wants to resell a vehicle, some loan programs have a resale value guarantee provision, which provides downside protection on a vehicle's value.

Technology

Tesla develops many components in-house, such as batteries, motors, and software.

Tesla's technology strategy focuses on pure-electric propulsion technology and transferring other approaches from the technology industry to transportation, such as online software updates. Tesla allows anyone in good faith to use its technology patents. Licensing agreements include provisions whereby the recipient agrees not to file patent suits against Tesla or copy its designs directly. Tesla retained control of its other intellectual property, such as trademarks and trade secrets, to prevent direct copying of its technology.

Batteries: Unlike other automakers, Tesla does not use individual large battery cells but thousands of small, cylindrical, lithium-ion commodity cells like those used in consumer electronics. It uses a version of these cells designed to be cheaper to manufacture and lighter than normal cells by removing some safety features. According to Tesla, these features are redundant because of the advanced thermal management system and an intumescent chemical in the battery to prevent fires. Panasonic is the sole supplier of the cells for Model S, Model X, and Model 3 and cooperates with Tesla in the Giga Nevada. As of January 2021, Panasonic can produce 39 GWh per year of batteries at Giga Nevada. Panasonic and Contemporary Amperex Technology (CATL) supply Tesla's battery cells used in Giga Shanghai. These batteries are more traditional prismatic (rectangular) cells used by other automakers.

During Tesla's Battery Day event on September 22, 2020, Tesla announced the next generation of batteries, featuring a tab-less battery design to increase the range and decrease the price of Tesla vehicles. The new battery is named the "4680", and its dimensions are 46 millimeters (1.8 in) wide by 80 millimeters (3.1 in) tall. Tesla predicted that the new batteries could be up to 56% cheaper and allow the cars to have a 54% longer range. A more efficient production process, new battery design, cheaper anode and cathode resources, and better vehicle integration will help the company to achieve its goals.

Motors: Tesla makes two kinds of electric motors. Its oldest design in production is a three-phase four-pole alternating current induction motor with a copper rotor (which inspired the Tesla logo).

This motor acts as the rear motor in the Model S and Model X. The Model 3 and the Model Y use newer, high-efficiency permanent magnet motors. The permanent magnet motors increase efficiency, especially in stop-start driving. Tesla Semi will use the front engine of 2019-onward versions of the Model S and X.

Autopilot: Autopilot is an advanced driver-assistance system developed by Tesla. The system requires active driver supervision at all times.

Since September 2014, all Tesla cars have been shipped with sensors (initially hardware version 1 or "HW1") and software to support Autopilot. Tesla upgraded its sensors and software in October 2016 ("HW2") to support full self-driving in the future. HW2 includes eight cameras, twelve ultrasonic sensors, and forward-facing radar.

HW2.5 came into existence in mid-2017 and upgraded HW2 with a second graphics-processing unit (GPU) and a driver-facing camera for the Model 3 only. HW3 was released in early 2019 with an updated and more powerful computer, employing a custom Tesla-designed system on a chip.

Full self-driving: Full Self-Driving (FSD) is an optional extension of Autopilot promoted as eventually being able to perform fully autonomous driving. Tesla's approach to achieving full autonomy differs from that of other companies. Waymo, Cruise, and other companies rely on highly detailed (centimeter-scale) three-dimensional maps, lidar, cameras, and radar and ultrasonic sensors in their autonomous vehicles. Tesla's approach uses coarse-grained two-dimensional maps, cameras (no LIDAR), radar, and ultrasonic sensors. Tesla claims that although its approach is much more complex, it will ultimately be more helpful because its vehicles can self-drive without geofencing concerns.

Tesla's self-driving software went through several pieces of training on over 20 billion miles driven by Tesla vehicles as of January 2021. But most experts believe that Tesla's approach of achieving full self-driving by eschewing LIDAR and high-definition maps is not feasible.

11

LEGACY AUTOMAKERS IN ELECTRIC VEHICLES

"I believe the auto industry will change more in the next 5 to 10 years than it has in the last 50."

—Mary Barra

(CEO & Chairman, General Motors)

With the governments of one country after another announcing bans on fossil fuel vehicles in the recent past, the global auto majors are finally waking up and smelling the coffee.

These automakers aren't working on electrification solely for innovation's sake. These EV plans come in the face of two threats to internal combustion engines: the rise of Tesla and a growing movement to restrict the use of traditional combustion engines to cut down on vehicle emissions.

Stellantis, the world's fifth-largest automaker, plans to sell only electric cars in Europe by 2030. Toyota expects to release about 30 electric models in the same timeframe. GM hopes to stop making cars with internal combustion engines by 2035.

In August 2022, California announced that from 2035, all new cars sold in the state – the most populous in America – must be zero-emission. The European Union has also taken steps to ban the sale of fossil-fuelled vehicles – and even hybrids – by 2035. China wants at least half of all new cars to be electric, plug-in hybrid, or hydrogen-powered by that time.

The race is toward the electrification of cars by 2035 – which is about 12 years away from now. Not a long time in the world of automobiles.

Automakers EV Plans

In the first edition of this book, I mentioned that "as of January 18, 2019, an investment tsunami in electric cars and batteries by global automakers now totals $90 billion".

Well, just over three years later, that tsunami looks like a ripple in the water.

The most recent analysis shows carmakers planning to spend an estimated $515 billion over the next five to 10 years to develop and build new battery-powered vehicles and shift away from combustion engines. That's more than half a trillion dollars!

The investment is led by German automaker Volkswagen, with a planned investment of $112 billion through 2030. Chinese automakers will account for $124 billion as investments over the next decade, and U.S. automakers GM and Ford expect to spend nearly $60 billion through 2025.

In April 2022, Honda announced it would invest approximately 5 trillion yen ($40 billion roughly) to electrify its lineup over the next decade.

These investments do not include the tens of billions of dollars invested in additional production capacity by the world's largest battery companies, many in cooperation with their automaker partners.

Meanwhile, political and regulatory pressure is building on the world's carmakers to begin phasing out production of fossil-fueled vehicles, including gasoline-electric hybrids, over the next 10-15 years while ramping up output of fully electric models.

Several countries, from Singapore to Sweden, have said they will ban sales of new combustion engine vehicles by 2030. U.S. President Joseph Biden has said he wants 40% to 50% of sales to be electric vehicles by 2030.

There's no longer any doubt that the auto industry is going electric. Every international automaker is introducing battery cars, and plug-in hybrid options are increasingly part of model lines. But some auto companies have gone further and set dates for when they will produce only "electrified" cars (batteries and hybrids) and pure EVs. Others have set dates specific when they will stop developing internal-combustion engines, and some have already stopped.

Let's look at how the biggies of automobiles are chalking out their vehicle electrification strategies.

Volkswagen

First and foremost – Volkswagen will be spending a mammoth $100 billion on developing EVs.

The German automaker announced in December 2021 that more than half the money Volkswagen Group spends in the next five years would go towards electric vehicles and other high-technology efforts.

By the end of 2026, the company expects that 25% of the vehicles it sells globally will be electric. Electric vehicles account for 5% to 6% of the company's sales.

As it prepares for that shift, Volkswagen will spend €89 billion, the equivalent of a little more than $100 billion, to speed its push towards electrification. This investment marks the first time the company is budgeting more than half its spending on EVs, roughly $180 billion through 2026. At the same time, expenditures on hybrid vehicles, which use batteries along with gasoline engines, will be reduced by 30%.

The expenditures will include new factories, such as a new plant at the company's headquarters in Wolfsburg, Germany. Volkswagen's Project Trinity, a new flagship electric car planned for 2025, will be built at Wolfsburg. VW also outlined plans for several new facilities throughout Europe.

The company envisions a path in which Wolfsburg gradually becomes a self-contained electric vehicle factory, the hub of a global EV empire. Instead of fitting gas tanks and engines to car frames, workers will be slotting in battery cells. Millions of the company's vehicles will be all-electric by 2030, and by mid-century, few fossil-fuel cars, if any, will be made here.

Automobile companies like Volkswagen, Audi, Bentley, Bugatti, Cupra, Jetta, Lamborghini, Porsche, SEAT, Skoda, Volkswagen Commercial Vehicles, and the mobility service provider Moia also confirm that they plan a total renewal of their European production activities. The focus is on BEV production at its Wolfsburg and Hanover manufacturing sites in the German state of Lower Saxony.

Volkswagen Group CEO Herbert Diess confirmed that the facilities earmark the production of the Volkswagen brand's new flagship BEV, known under the internal working title Project Trinity. They will also provide excess production capacity for ID.3, which is currently producing alongside ID.4, ID.5, Audi Q4 SUV, Audi Q4 Sportback, and SEAT el-Born at Volkswagen's Zwickau factory.

Volkswagen is also planning to create a separate battery manufacturing company. According to VW, the battery business is expected to generate up to €30 billion in revenue, more than the company's Skoda passenger car division currently develops.

Under the Volkswagen Group's plans, its Salzgitter plant will shift activities from ICE production to become home to a new European battery production hub, where its planned unified battery cells will get assembled in 2025.

It is expanding its manufacturing outside its German base also. Volkswagen and partners have announced plans to invest €10 billion into producing electric vehicles and batteries in Spain – €3 billion more than previously promised.

VW also announced a partnership deal with Iberdrola, Spain's largest power utility. The pair aim to build a solar park that will partly power the battery plant, which will come up in Sagunto, near Valencia.

The German automaker is taking the fight to the U.S., a market where Tesla is the overwhelmingly dominant EV manufacturer. VW announced plans in March 2022 to invest $7.1 billion in North America, where it will add new production capacity for all-electric models like the ID. Buzz microbus.

Ford

Ford has announced bold plans to step into an all-electric future with an investment outlay of $50 billion between 2022 and 2026. It will run its EV unit separately from its legacy combustion engine business. Christened Ford Model e, the EV business unit, plans to build 2 million EVs in 2026, about one-third of its annual global production. The EVs will contribute 50% of Ford's total volume by 2030.

Ford's reorganization is one of the most sweeping steps a traditional automaker takes in preparation for the transition to electric vehicles.

Doug Field, previously a top engineer for both Tesla and Apple, leads Model e's product development as its chief EV and digital systems officer. He's also in charge of developing software and embedded systems for Ford.

Ford has already received an overwhelming response to its all-electric model lineup currently in the market. The Ford F-150 Lightning stopped taking reservations in December 2021 after it received nearly 200,000 bookings – far more than its current production capacity. The production for this model will be ramped up to 80,000 units annually by 2024.

Mustang Mach-E sales crossed 21,000 units in 2021. The electric van, E-transit, also received an encouraging response from the market and has booked over 10,000 orders.

Ford has confirmed a fully electric Puma will join the range alongside a pair of new crossover EVs by 2024 as it rolls out nine electric vehicles within two years. It has also pledged to go carbon-neutral by 2035.

General Motors

Ford's archrival, General Motors (GM), has also announced aggressive plans for an all-electric future. In 2021, GM said it would boost its EV spending to $35 billion by 2025. In early 2022, General Motors CEO Mary Barra announced Tuesday that GM aims to deliver 400,000 EVs in North America by the end of 2023.

GM plans to release 20 EVs in the United States through 2025, with ten additional EVs for global markets. The company plans an aggressive product expansion. The GMC Hummer EV is already in the delivery stage, while Silverado EV production will start in 2023. Chevy is also planning electric versions of the Equinox and Blazer crossovers, GMC will have an EV version of the Sierra, and GM is building BrightDrop vans for commercial customers. Cadillac will launch its Lyriq crossover this spring, with the flagship Celestiq arriving at some point down the line. Buick will also gain an EV in the coming years, though an announcement related to the same is not official till now.

But the future will bring truly affordable GM electric vehicles, thanks tothe partnership with Honda. Those lower-priced models, including compact crossovers, are expected to sell in 2027.

The automaker also announced it would boost the production of battery cells in the US and speed up plans to build two new battery plants. This plant is in addition to battery plants already under construction in Tennessee and Ohio.

Toyota

Toyota has been a reluctant giant to join the battery-electric car movement. But the way the world is moving, it has no choice.

For 20 years, Toyota rejected the very idea of EVs. The world's most profitable giant automaker argued that gasoline hybrids would be the best and most practical way to slash vehicle emissions. Until then, its hydrogen fuel-cell cars were ready for primetime in 2030.

But then Tesla came and changed the rules of the game, and Toyota also made a change in its strategies.

In mid-December 2021, Toyota CEO Akio Toyoda presented his company's new and expanded plans to boost its production of battery-electric vehicles. For the sake of summary, there were *a lot* of big-production and big-dollar promises. In the 25-minute media briefing, Toyoda set a goal of 3.5 million battery-electric vehicles a year by 2030 (out of Toyota's 10 million total globally), through no fewer than 30 different Toyota and Lexus models in all market segments. And in total, he pledged a whopping $70 billion toward electrification. Of that amount, approximately $35 billion will be invested in BEVs starting in 2022 through 2030.

In addition to continuing to develop hybrids (HEVs), plug-in hybrids (PHEVs), and fuel-cell electric vehicles (FCEVs), Toyota will expand its offering of battery electric vehicles (BEVs)

Toyota's commitment is notable. The company has been the most significant automaker fighting what many now see as an inevitable transition from gasoline engines to battery-powered vehicles. It is finally a recognition that EVs will be a substantial part of most of its markets, at varying rates, and will happen this decade.

Akio Toyoda presented five "Beyond Zero" vehicles in front of a row, including the 2022 Toyota bZ4X compact SUV. The other four included a much sleeker "coupe SUV," a small SUV for Europe and Japan, a midsize sedan (think Camry), and a large SUV with available third-row seating (think Highlander). Toyota will stick with the bZ branding. Beyond Zero—for other BEVs, including a compact SUV and a small crossover. Toyoda describes it as "a small battery EV with a comfortable interior designed with Europe and Japan in mind." Toyota also unveiled 11 more vehicles for 16 EV concepts from Toyota and its Lexus luxury brand.

While Toyota will have EVs in several segments, Lexus will become the company's EV spearhead. Toyota's luxury brand aims for all its sales in Europe, North America, and China to be battery-electric vehicles by 2030—just eight years from now. And it expects 100 percent of its global sales to be EVs five years after that, in 2035—or only 23 years behind Tesla.

Now, let's move on to the other well-known Japanese automaker – Honda.

Honda

Honda, the Japanese automaker, will sell only EVs and hybrids in Europe after 2022. By 2030, Honda says 40 percent of its North American vehicle sales will be either battery electric or hydrogen. By 2040, all gas cars will be moving out. Honda's first two EVs (one Honda SUV called the Prologue, one Acura). GM will build in 2024, but according to Dave Gardner, a North American vice president, "We intend to produce in our factories, and we intend to utilize that resource." Honda is one of many automakers pushing hydrogen (fuel cells) instead of batteries as fuel sources. The company has been building Clarity fuel cell electric vehicles (FCEVs) since 2008; the Clarity name also applies to traditional EVs and plug-ins.

Like every traditional automaker, Honda is looking down the road and seeing an end to selling good old-fashioned gasoline-powered cars.

In April 2022, Honda CEO Toshihiro Mibe revealed his company's plans, which call for launching 30 electric car models between now and 2030, with a target of 2 million sales a year. The company is preparing to drop a $64 billion wad of cash to achieve the milestone and on several investments designed to keep it competitive in the coming EV Age.

Over the next ten years, an investment of $ 40 billion will be in electrification. That includes expanding Honda's EV platforms, but some investment will help create new growth opportunities in space exploration, eVTOLs, avatar robots, and more. The company is also planning to spend an additional $80 million a year on startups that could help expand its business and shift from selling products alone to offering combined solutions. Mibe said Honda would also actively pursue inter-industry collaboration and alliances like the ones it announced already with General Motors and Sony.

As part of his presentation, Mibe said Honda is planning to introduce an insanely cheap mini-EV model for commercial use in Japan that will sell for around $8,000, and that vehicle will be on sale from 2024. Honda also plans to introduce two mid to large-size EVs - a Honda Prologue SUV and an Acura SUV - which are currently under development with General Motors at prices that are supposed to be cost-competitive with ICE vehicles.

Honda had previously committed to launching ten new EV models in China under its e: N Series by 2027, with two models available to customers in 2022. It plans to build a dedicated EV plant in Guangzhou and Wuhan to support production in one of its most important markets.

While Honda will utilize GM's Ultium architecture and EV platform for some electric models, it fully intends to chart its course. The company plans to introduce the Honda e: Architecture in 2026, an EV platform that includes hardware and cloud-connected software.

Like many other automakers, Honda sees the potential for software-defined vehicles to help generate recurring revenue through third-party applications. To make that happen, it will build an application layer continuously updated over the air on top of the vehicle operating system.

Despite its previous public skepticism of prospective BEV sales, Honda does plan to deliver its first all-electric volume model in the Prologue, which it has co-developed with GM. This all-electric SUV comes from GM's Ultium electric platform and will arrive in 2024 alongside an all-electric Acura SUV.

The critical challenge in the EV era is the global procurement of batteries, and Honda has two basic approaches for its battery procurement strategy.

For the time being, Honda will ensure stable procurement of liquid lithium-ion batteries in each region by strengthening the external partnership. In North America, Honda will procure Ultium batteries from GM. Aside from GM, Honda is exploring the possibility of creating a joint venture company for battery production. In China, it will further strengthen the collaboration with CATL. And in Japan, it will procure batteries for mini-EVs from Envision AESC.

From the second half of the 2020s, Honda will further accelerate its independent research and development of next-generation batteries. For the all-solid-state batteries it currently has under development, Honda has decided to build a demonstration line, investing approximately 43 billion yen, intending to make it operational in Spring 2024. Honda aims to adopt its next-generation batteries and introduce them to the market in the second half of the 2020s.

Renault-Nissan-Mitsubishi Alliance

In January 2022, Renault and Nissan announced that they would work more closely to make electric cars. They detailed a $26 billion investment plan over the next five years to stay competitive in the switch to cleaner driving.

The two-decade-old alliance, including Mitsubishi Motors, plans to increase the number of common platforms for electric vehicles (EV) to five from four. They will build a combined EV line-up of 35 cars by 2030, and by 2026 four-fifths of their models will share common platforms, up from 60% now.

Japan's No. 3 carmaker plans to launch 23 electrified vehicles by 2030, including 15 pure EVs. It also wants to reduce lithium-ion battery costs by 65% within eight years and introduce potentially game-changing all-solid-state batteries by March 2029.

To power the new EVs, the alliance plans to secure 220 gigawatts (GWh) hours of battery production capacity by 2030, providing a more significant scale that would allow them to halve battery costs by 2026 and reduce them by 65% by 2028. Further detail on the same is still pending.

Nissan aims to launch an EV with its proprietary all-solid-state batteries (ASSB) by the fiscal year 2028 and ready a pilot plant in Yokohama as early as the fiscal year 2024. By reducing charging time to one-third, ASSBs will make EVs more efficient and accessible. Further, Nissan expects ASSB to bring the cost of battery packs down to $75 per kWh by the fiscal year 2028 and aims to bring it further down to $65 per kWh to achieve cost parity between EV and gasoline vehicles in the future.

The Alliance has offered 10 EV models built on four EV platforms, including CMF-EV, the global multivalent electric platform, and CMF-AEV, the most affordable EV platform. The coming years will see a complete development of these platforms.

Firstly, the Alliance will add a new CMF-BEV platform in 2024, designed to be the world's most competitive compact electric platform. It will provide up to 400 km range and outstanding aerodynamic performance, helping to reduce cost by 33% and power consumption by more than 10%. Nissan has selected this new platform for its pure-EV successor of Micra in Europe. Secondly, the existing CMF-EV platform will evolve from two models today to more than 15 by 2030, spanning five brands and 1.5 million cars per year.

Renault has said it will become an electric-only brand in Europe by 2030, joining Ford of Europe, Stellantis brands, and others in dropping combustion engines pledge as the European Union pushes to halt internal combustion sales by 2035.

Mercedes-Benz

Mercedes-Benz is getting ready to go all-electric by the end of the decade when market conditions allow. Shifting from electric-first to electric-only, the world's pre-eminent luxury car company is accelerating toward an emissions-free and software-driven future. In total, investments into battery electric vehicles between 2022 and 2030 will amount to over €40 billion.

By the end of 2022, Mercedes-Benz will have battery electric vehicles (BEV) in all segments the company serves.

From 2025 onwards, all newly launched vehicle architectures will be electric-only, and customers will be able to choose an all-electric alternative for every model the company makes. These electric architectures will be :

1. MB.EA will cover all medium to large-size passenger cars, establishing a scalable modular system as the electric backbone for the future EV portfolio.
2. AMG.EA will be a dedicated performance electric vehicle platform addressing technology and performance-oriented Mercedes-AMG customers.
3. VAN.EA – will make electric vans and Light Commercial Vehicles, contributing to future emission-free transportation and cities.

Mercedes-Benz is preparing its global production network for electric-only output with the pace of the ramp-up designed to follow market demand. This year, Mercedes-Benz will produce eight electric vehicles at seven locations on three continents.

To enhance manufacturing efficiency, Mercedes-Benz is joining forces with GROB, a German global leader in highly innovative battery production and automation systems, strengthening its battery production capacity and know-how. The cooperation focuses on battery module assembly as well as pack assembly. Mercedes-Benz also plans to install a new battery-recycling factory in Kuppenheim, Germany, to develop and secure recycling capacity and know-how.

Mercedes-Benz will need a battery capacity of more than 200 Gigawatt hours and plans to set up eight Gigafactories for producing cells with its partners worldwide. This decision is in addition to the already planned network of nine plants dedicated to building battery systems. Next-generation batteries will be highly standardized and suitable for more than 90% of all Mercedes-Benz cars and vans while being flexible enough to offer individual solutions to all customers.

Mercedes-Benz is also working on setting new standards in charging: "Plug & Charge" will allow customers to plug-in, charge, and unplug without extra steps needed for authentication and payment processing. Plug & Charge will go live with the market launch of the EQS later this year. Mercedes Charge is already one of the world's largest charging networks and currently comprises more than 530,000 AC and DC charging points worldwide. Furthermore, Mercedes-Benz is working with Shell on expanding the charging network. Customers will get enhanced access to Shell's Recharge network of over 30,000 charge points by 2025 in Europe, China, and North America – including over 10,000 high-power chargers globally. Mercedes-Benz is also planning to launch several premium-charging sites in Europe, offering a bespoke charging experience with top-notch facilities.

Audi

Audi, the company with four rings as its logo, is a well-established German automaker and a subsidiary of Volkswagen Group. Audi has announced goals to become a sustainable, social, and technological leader in its continued journey toward vehicle electrification by 2030. This announcement is part of a new corporate strategy from Audi called "Vorsprung 2030," aiming to accelerate the automaker's electrified transformation slotted initially for 2026. Furthermore, Audi will continue to phase out internal combustion engine (ICE) vehicle production, approaching the end date of 2033.

Audi's goal is to offer more than 30 electrified cars by 2025, with 20 being full BEVs. They expect electrified vehicles to cover 40% of all Audis sold by 2025. It will use four platforms to explode its range of electric cars in the decade: MEB, MLBevo, J1, and PPE (Premium Platform Electric).

Another primary focus will be developing an ecosystem for its electric fleet that includes autonomous driving and seamless updates, giving future Audis an unmistakable DNA. Volkswagen Group's software arm, CARIAD, will develop a scalable software platform with a standardized operating system and cloud connectivity for all its brands by 2025.

Hyundai

In early March 2022, Hyundai Motor Company unveiled a strategic roadmap to accelerate its electrification ambition as it pursues sustainable progress for the company. Under the plan, the company aims to boost annual global BEV sales to 1.87 million units and secure a 7 percent global market share by 2030.

Hyundai plans to introduce 17 BEV models, 11 Hyundai marque models, and six models from Genesis luxury brands by 2030 as it seeks to expand the BEV spectrum.

The new Hyundai BEV models will include three sedan models, six SUVs, one light commercial vehicle, and one new type model. This year, Hyundai began sales of IONIQ 6, followed by IONIQ 7 in 2024.

Genesis luxury brand, the BEV line-up consists of two passenger cars and four SUVs, including the Electrified GV70, launching this year. At the start of 2025, Genesis will electrify all the newly launched models.

Beyond existing BEV production facilities centered in Korea and the Czech Republic, Hyundai plans to gradually expand its BEV manufacturing bases, starting with an Indonesian plant that recently began operation.

Hyundai Motor plans to introduce an Integrated Modular Architecture (IMA), evolved from the electric global modular platform (E-GMP) that is the foundation of IONIQ 5 and GV60, successfully launched in 2021. The IMA will utilize it as Hyundai Motor's passenger BEV platform and its exclusive purpose-built vehicle (PBV) platform, helping streamline production processes and reduce costs.

Unlike the existing BEV development system, which has different battery packs for each model, IMA can be equipped with standardized battery packs to attach flexibly regardless of the model to improve cost efficiency. The new architecture can secure sufficient energy density and shorten charging time through the cell-to-pack system.

Hyundai will set five types of standardized batteries on the IMA's, according to model needs.

Hyundai Motor Company announced that it intends to "increase the local procurement rate of batteries through strategic alliances with battery companies in major regions" (including the US). One of the directions is diversification.

According to a press release, a sufficient battery supply is secured through 2023, while by 2030, the Hyundai/Genesis brands will need 170 GWh annually. On average, that would be over 90 kWh per battery pack (assuming 1.87 million BEVs).

Hyundai is working with partners on next-generation batteries. As per expectation, more than 50% of the supply of batteries will come through partner channels.

Kia

Kia Corporation has followed its parent company, Hyundai Motor Group, by outlining its roadmap to further sustainability by 2030. The strategy includes 14 BEV models in the Kia line-up by 2027 and an expected 1.2 million BEV sales annually by 2030. Additionally, Kia will deliver its dedicated purpose-built vehicles starting in 2025.

Kia plans to launch at least two BEV models per year beginning in 2023, striving toward a complete lineup of 14 BEVs by 2027. Kia's previous plan was to deliver 11 models by 2026, but it will now add two electric pickup trucks. One will be a dedicated "strategic model" for emerging markets, the other an entry-level BEV.

To support this slew of BEVs in its pipeline, Kia is investing in its battery technology to support the growing demand. Kia expects this demand to increase from 13 GWh to 119 GWh in 2030 due to increasing EV sales.

In addition to garnering batteries from an Indonesian joint venture, Kia said it would continue outsourcing cells to global battery manufacturers. Kia also said it plans to increase battery energy density by 50% by the decade's end and reduce system costs by 40%.

Hyundai Motors and Kia plan to invest 21 trillion won ($16.5 billion) to boost the production of electric vehicles in South Korea. The project also includes establishing a new factory that would ultimately have the capacity to make around 150,000 cleaner cars a year.

Under the plan, the two automakers aim to increase annual EV production in the nation to 1.44 million units by 2030 from an expected 350,000 units this year. That forecast 1.44 million output would account for about 45% of Hyundai and Kia's planned global EV production volume by then.

BMW

BMW, the automaker from Bavaria, is a holdout among German carmakers in terms of fully electric vehicles, as opposed to here-and-now plug-ins.

BMW stands out among automakers for its reluctance to name a date for the phase-out of the combustion engine and a cautious approach to the rollout of electric models. Unlike Mercedes, which has pledged to be all-electric by 2030 "where market conditions allow," executives at BMW believe combustion engine models will still make up half of all sales by the end of the decade, ever stricter EU emissions rules notwithstanding.

The time will show how the electric market will develop. Establishing a sufficient charging infrastructure to attract average consumers for emission-free cars will be difficult. Therefore,

BMW will produce electric, hybrid, and combustion engine cars on the same assembly lines by the same staff.

But BMW is undoubtedly BEV-electrifying, too. In March 2022, BMW CEO Oliver Zipse said, "By the end of 2025, we will have delivered around two million fully electric vehicles to customers. We will also be growing our sales of fully electric models by well over 50% per year over the next few years—more than ten times the figure for 2020."

By 2030, the company said approximately half its global sales would be battery cars. The electric iX SUV and i4 sedan are both market-bound now. The high-end European makers—Audi, BMW, Mercedes-Benz, and Porsche—justified high prices with high levels of performance and technology. Now technology encompasses zero-emissions performance.

BMW will enter the next chapter of its electrification strategy with the launch of the Neu Klasse (New Class) platform. The architecture is currently under development. As per expectation, the fruition of this process is in 2025 in the form of a fully-electric 3 Series. The initial focus of this new platform will be models from the midsize premium segment, and vehicles on the new architecture will only have a purely electric powertrain.

Following its initial launch, the Munich based manufacturer will expand the scope of the Neue Klasse for use in electric vehicles from other segments. There will be a full range of zero-emissions cars based on the architecture, ranging "from the high-volume segment to exclusive high-performance models." Meanwhile, the company's combustion-fuelled models will continue on other platforms.

While the company is using the 5th generation eDrive system on current models like the i7, iX, and the i4, the goal for BMW in 2025 is to launch a new cluster architecture that can deliver the entire BMW product portfolio with one building principle. The Neue Klasse will come with the new Generation 6 drivetrain generation for the first time with more output, unique cell chemistry, and new cell formats.

The Neue Klasse BMW models will use a newly developed electric drivetrain and next-gen battery packs (Gen 6). An electric drive train based on the hydrogen fuel cell could also be a part of this.

The Gen 6 battery packs of the Neue Klasse platform-based BMW EVs will consist of cells with significantly higher energy density. BMW Group is also working on commercializing solid-state battery (All-Solid-State-Battery) technology this decade.

BMW Group plans to substantially reduce the use of primary material with the switch to the new battery technology. The company wants the next-gen electric powertrain to offer a range comparable to state-of-the-art internal combustion engines, and that too at the same level of manufacturing costs.

BMW Group will start production of the BMW Neue Klasse models in 2025, marking the beginning of the third phase of transformation. Initially, the manufacturing will take place in a country offering lower production costs than Germany – Hungary. BMW Group plans to manufacture Neue Klasse models initially at a new factory in Debrecen, a city located in the eastern part of the country, starting in 2025. Hungary will be the production site of the next-gen BMW 3 Series and next-gen BMW X3.

The BMW Group Plant Munich will roll out Neue Klasse BMW electric cars starting in 2026. The BMW Group has confirmed these two factories for producing Neue Klasse models, and it plans to gradually transfer the new architecture to global production sites in the coming years.

Stellantis

Stellantis is setting the course for 100% of sales in Europe and 50% of sales in the United States to be battery electric vehicles (BEVs) by the end of this decade. It plans to offer more than 75 battery electric vehicle models and reach global annual BEV sales of 5 million units by 2030.

The Stellantis plan calls for serving a wide variety of vehicle segments and ranges with a combination of four state-of-the-art BEV-centric vehicle platforms. – STLA Small, Medium, Large, and Frame – three electric drive modules (EDMs) to provide modularity and performance and two battery cell chemistries to ensure affordability for all. Those combinations offer industry-leading features, driving ranges of between 500-800 kilometers/300-500 miles, and front-, rear- and all-wheel-drive propulsion.

The four platforms - 3 unibody and one body on the frame - are designed with a high level of flexibility (length and width) to offer the most efficient solution for each vehicle category, from city cars to pickup trucks and SUVs. This level of flexibility, in addition to component sharing across platforms, will reduce complexity and deliver economies of scale, with each platform capable of supporting the production of up to two million units per year.

Stellantis will utilize three different EDMs that cover the full global vehicle range from the STLA Small up to the STLA Frame platform. This third-generation technology is highly flexible, with a modular, scalable design (from 70kW to 330 kW) and a high level of commonality to optimize overall cost efficiency. A vital element of this strategy is the development of one power inverter for all three EDM families with scalable technology, a standard microprocessor, and in-house controls and software.

By 2024, the company also plans to use two battery chemistries: a nickel cobalt-free alternative, which offers a low, stable cost while at the same time optimizing density, and a high energy-density option which is nickel based.

These two chemistries have many synergies, including process and component materials. From 2026, they will share the same cell-to-pack installation, upgradeable to respond to customer needs and improve energy density or charging speed. In addition, the company's first competitive solid-state battery technology is targeted for introduction by 2026.

12

GLOBAL OUTLOOK FOR ELECTRIC VEHICLES

"Environmentally friendly cars will soon cease to be an option – they will become a necessity."

—Fujio Cho

(Honorary Chairman, Toyota)

Global EV sales reached 6.75 million units in 2021, 108 % more than in 2020. This volume includes passenger vehicles, light trucks, and light commercial vehicles. The global share of EVs (BEV & PHEV) in global light vehicle sales was 8.3 % compared to 4.2 % in 2020. BEVs stood for 71 % of total EV sales, PHEVs for 29 %.

In the first six months of 2022, 4.1 million new passenger electric vehicles went for registration globally. June 2022 witnessed selling more than 900,000 electric cars globally, which is 54% more than a year ago. That's a new massive record and a clear indication that we are heading towards 1+ million per month and 10+ million per year!

It's worth noting that the plug-in electric car market share increased to 16%, while all-electric cars achieved a 12% market share record for the first time.

We are clearly past the inflection point, and customers are increasingly choosing electric vehicles.

So, how will it pan out from here? Let’s have a look at the EV sales forecasts made by some reputed organizations across the world.

Bloomberg New Energy Finance

Bloomberg New Energy Finance (BNEF) has updated its outlook for how road transport could evolve over the next 30 years in its 7th annual Electric Vehicle Outlook (EVO) 2022. As per the report, there are now almost 20 million passenger EVs on the road, 1.3 million commercial EVs (buses, delivery vans & trucks), and over 280 million electric mopeds, scooters, motorcycles, and three-wheelers.

In the near-term outlook, BNEF forecasts that passenger EV sales will rise from 6.6 million units in 2021 to 20.6 million in 2025. This growth will represent 23% of new passenger vehicle sales globally, up from just under 10% in 2023.

The EV share in some markets will be much higher, with EVs holding a 39% share in China and Europe in 2025. Germany, the UK, and France will have EVs garnering a 40-50% share by that year. The EV adoption will be lopsided, with China and Europe accounting for 80% of EV sales in 2025.

Most importantly, the market is shifting from being driven primarily by policy to one where organic customer demand is the most critical factor. The acceleration in EV adoption means that ICE vehicle sales peaked globally in 2017 and are now in a 'permanent decline.' By 2025, BNEF expects that there will be 77 million passenger EVs globally, representing 6% of the fleet.

From 2026 onward, BNEF's outlook splits into two scenarios. Techno-economic trends and market forces primarily drive the Economic Transition Scenario (ETS). They also assume that no enactment of new policies or regulations will happen to create an impact on the market. The second scenario investigates what a potential route to net-zero emissions looks like for the road transport sector by 2050. This Net Zero Scenario (NZS) looks primarily at economics as deciding which drivetrain technologies could achieve the goal.

Under ETS, passenger EV sales continue to rise quickly as battery prices fall. By 2040, global passenger EV sales will reach an annual sales volume of over 70 million units, with Europe and China having a 90% share of new vehicle sales as EVs. The global share of EVs will cross the 70% line by 2040.

Despite the relatively rapid growth of EV sales, there are still 800 million ICE vehicles on the road in 2040 – more than half the global car parc.

EVs take longer to spread in India, Southeast Asia, and other countries like Mexico, Brazil, and Russia. The ETS trajectory suggests that a two-tiered global auto market is emerging and that the benefits of electrification are ready to accrue very unevenly between wealthy and emerging economies. Urgent action is needed to close this gap.

Electric bus sales already comprise 44% of new sales and 18% of the global fleet. However, 98% of the global e-bus fleet is still in China. By 2040, the global e-bus fleet will reach 1.75 million units, representing 62% of all buses on the road.

Buses and two-and three-wheeled vehicles will achieve the highest EV adoption rates by 2040 in the ETS, followed by passenger cars and light commercial vehicles.

Now let us look at BNEF's Net Zero Scenario (NZS). Zero-emission vehicles need to represent 61% of global new passenger sales by 2030, 93% by 2035, and 100% by 2038 to get back on track for a net-zero global fleet by 2050,

The fleet of electric passenger vehicles will hit 469 million units in 2035 in ETS but needs to jump to 612 million by the same date in NZS. Much of the gap will have to close in emerging economies, and the 'adoption gap' between the two scenarios is much smaller for two/three-wheelers and buses.

Electric vehicles represent a $9 trillion market opportunity between today and 2030, and $53 trillion between today and 2050, in the ETS. This opportunity increases to $82 trillion between today and 2050 in the NZS.

International Energy Agency (IEA)

The International Energy Agency (IEA) has also worked out its projections of EV volumes in its report for 2022 with two scenarios – The stated Policies Scenario and the Announced Pledges Scenario.

The Stated Policies Scenario reflects existing policies and measures, as well as policy ambitions and targets that governments worldwide have legislated. It includes current EV-related policies and regulations and future developments based on the expected impacts of announced deployments and plans from industry stakeholders.

The Announced Pledges Scenario assumes that the announced ambitions and targets made by governments worldwide, including the most recent ones, are met in full and on time. With regards to electro-mobility, it includes all current significant electrification announcements targets, longer-term net zero emissions, and other pledges, regardless of whether or not these have been a part of the legislation.

In the Stated Policies Scenario, the global EV stock across all road transport modes excluding two/three-wheelers) is expanding rapidly from 18 million in 2021 to 200 million vehicles by 2030, an average annual growth of over 30%. In this scenario, EVs will account for about 10% of the road vehicle fleet by 2030. Total EV sales will reach 18 million in 2025 and over 30 million vehicles in 2030, representing 13% and over 20% of all road vehicle sales. This scenario is in comparison to almost 7 million in 2021.

In the Announced Pledges Scenario, based on the targets and pledges that go beyond stated policies, the global EV stock will reach over 85 million vehicles in 2025 and 270 million in 2030 (excluding two/three-wheelers). The share of EVs in the stock will get 14% in 2030. EV sales in this scenario will reach over 45 million vehicles in 2030, achieving a sales share of 33%.

For comparison, in the Net Zero Scenario, the global EV stock will reach over 100 million vehicles in 2025 and 350 million in 2030 (excluding two/three-wheelers). The share of EVs in the stock will get 20% in 2030. In 2030, EV sales will reach over 65 million vehicles, representing a sales share of almost 60%. This sales share is 80% higher than the Announced Pledges Scenario, indicating that government pledges fall short of reaching net zero by 2050.

In the Stated Policies Scenario, sales of electric Light Duty Vehicles (LDVs - passenger cars and light commercial vehicles) will rise from almost 7 million in 2021 to over 17 million in 2025 (a sales share of about 15%) and over 30 million in 2030 (more than 20%). In the Announced Pledges Scenario, sales of electric LDVs will reach about 45 million in 2030 (a sales share of 35%), 50% above the Stated Policies Scenario.

In the Net Zero Scenario, the sales share of electric LDVs is more than 60% in 2030, which is over one-and-a-half times higher than in the Stated Policies Scenario and 80% higher than in the Announced Pledges Scenario.

Electric two/three-wheelers will continue to be the largest EV fleet among all road transport modes. Asia, where two/three-wheelers are prevalent today, is the primary growth center. The global stock of electric two/three-wheelers in the Stated Policies Scenario increases from over 35 million in 2021 to 245 million in 2030. This increase will account for over a quarter of the total stock in 2030. The electric two/three-wheeler sales will increase from 10 million in 2021 to 40 million in 2030, accounting for more than half of all sales.

In the Announced Pledges Scenario, the global stock of electric two/three-wheelers rises to over 330 million in 2030, 35% of the total inventory for two/three-wheelers. This growth corresponds to sales close to 55 million in 2030, amounting to about 65% of all sales.

In the Net Zero Scenario, electric two/three-wheelers reach a sales share of 85% in 2030, 60% and 30% higher than the Stated Policies and Announced Pledged scenarios.

Announced Country / Region Targets for EV deployment

Many countries across the world have announced vital policies and measures for the deployment of EVs. The legislature is making legally binding commitments such as regulations and standards. The roadmaps with specific targets and ambitions that support the growth of electrification are underway.

Here are some of the significant announcements made in various parts of the world :

Country	Key policy measures and targets	Year announced
European Union	• Target: 13 million-passenger ZEV stock by 2025. • Voluntary ZEV targets: 15% share of car sales by 2025 and 35% by 2030, 15% share of van sales by 2025, and 30% by 2030 by vehicle manufacturers. • Ambition: at least 30 million-passenger ZEV stock by 2030 and nearly all passenger LDV and heavy commercial vehicle stock by 2050. • Ambition: 1 million publicly accessible charging stations by 2025 and 3 million by 2030.	• 2019 • 2019 • 2020 • 2020
Denmark	• Ambition: end the sale of new petrol, and diesel cars from 2030.No PHEVs will be on sale from 2035. Purpose: 100% share of ZEVs in urban bus procurements by 2025 and 100% ZEV urban bus fleet by 2030. • Ambition: 30% of ZEVs in new truck and bus sales by 2030, 100% by 2040.	• 2018 • 2021
France	• Target: 500 000 passenger PHEVs, 660 000 passenger BEVs and FCEVs, and 170 000 light commercial BEV and FCEV stock by 2023. Mark: 1.8 million passenger PHEVs, 3 million passenger BEVs and FCEVs, and 500 000 light commercial BEVs and FCEV stores by 2028.	• 2020
Germany	• Ambition: 50% share of electric urban buses by 2030. • Ambition: 15 million EVs on the road by 2030.	• 2020 • 2021
Norway	• Target: 100% share of ZEVs in passenger LDV sales by 2025. • Target: 100% share of ZEVs (or biogas) in urban bus sales by 2025. Target: 75% share of ZEVs in long-distance bus sales, 50% ZEVs in truck sales, and 100% share of ZEVs in heavy van sales by 2030.	• 2016 • 2016
United Kingdom	• Ambition: phase out petrol and diesel passenger LDV sales by 2030. All sales of passenger LDVs are to be BEVs or FCEVs by 2035. • Ambition: 30% of ZEVs in new truck and bus sales by 2030, 100% by 2040. • Ambition: phase out selling new mini petrol and diesel trucks by 2035 and larger trucks (>26 tonnes) by 2040.	• 2020 • 2021 • 2021
China	• Target: 20% share of NEVs in LDV and HDV sales by 2025. • Ambition: 100% share of EVs in passenger LDV sales by 2035 (of which 50% are NEVs and 95% of those are BEVs). Purpose: 100 000 FCEV sales in LDVs and HDVs by 2025 and around 1 million stock in the 2030-35 period. • Ambition: charging infrastructure sufficient to meet the needs of more than 20 million NEVs by 2025.	• 2020 • 2020 • 2022
India	• Ambition: 30% share of EVs in passenger LDV	• 2017

	sales by 2030. • Ambition: charging stations every 40 - 60 km on national highways or 700 charging stations by 2023 covering 35 000 - 40 000 km of national roads.	• 2021
Japan	• Target: 20 - 30% share of BEVs and PHEVs, 30 - 40% share of HEVs, and 3% share of FCEVs in passenger LDV sales by 2030. • Ambition: 100% electrified vehicles in passenger LDV sales by 2035.	• 2018 • 2020
Korea	• Target: 1.13 million passenger BEV and 200,000 passenger FCEV stock by 2025. • Ambition: 51% share of EVs in new vehicle sales by 2025 and 83% by 2030.	• 2020 • 2021
Canada	• Target: 20% ZEV LDV sales by 2026, 60% by 2030 and 100% by 2035. • Ambition: 30% of ZEVs in new truck and bus sales by 2030, 100% by 2040.	• 2021 • 2021
USA	• Ambition: 50% share of EVs in passenger LDV sales by 2030. • Ambition: 30% share of ZEV in all new medium- and heavy-duty commercial vehicle sales by 2030 and 100% by 2050 in 15 states and the District of Columbia.	• 2021 • 2020
Chile	• Ambition: 100% share of ZEVs in light and medium vehicle sales by 2035. Purpose: 100% share of EVs in public transport vehicle sales (buses, taxis, and shared taxis) by 2035. Ambition: 100% share of ZEVs in freight transport and intercity bus sales by 2045.	• 2021

Electric Vehicle Initiative (EVI)

The Electric Vehicles Initiative (EVI) is a multi-government policy forum dedicated to accelerating the introduction and adoption of electric vehicles worldwide. EVI was one of several initiatives in 2010 under the Clean Energy Ministerial (CEM), a high-level dialogue among Energy Ministers from the world's major economies.

Sixteen countries participate in EVI: Canada, Chile, China, Finland, France, Germany, India, Japan, the Netherlands, New Zealand, Norway, Poland, Portugal, Sweden, the United Kingdom, and the United States. The International Energy Agency (IEA) is the EVI Co-Ordinator.

The Clean Energy Ministerial (CEM) announced a new campaign in 2017 called EV 30@30 to speed up the deployment of electric vehicles and target at least 30 percent of recent electric vehicle sales by 2030. The campaign currently totals 13 member countries and 23 supporting companies and organizations.

The campaign supports the market for electric passenger cars, light commercial vans, buses, and trucks (including battery-electric, plug-in hybrid, and fuel-cell vehicle types). It also works towards deploying charging infrastructure to supply sufficient power to the vehicles deployed.

The 12th Clean Energy Ministerial in June 2021 led to the EVI Call to Action launch.

It is a call to close the gap between projected levels of EV sales in 2030 based on existing policies and the global sales needed to be on track with the Paris Agreement's objectives. Various governments and the private sector stand behind the Call through 9 practical actions that require implementation over the next two years.

The EVI Global EV Pilot City Programme (EVI-PCP) aims to create a global platform to facilitate communications and cooperation among leading global cities interested in stimulating and increasing the uptake of electric mobility within their jurisdictions.

13

CHINA

"God created the world…and the rest was made in China."
— Anonymous

The Chinese government uses the term "New Energy Vehicles (NEVs)" to designate plug-in electric vehicles eligible for public subsidies. It includes only battery-electric vehicles (BEVs) and plug-in hybrid electric vehicles (PHEVs).

In 2021, China recorded the most substantial growth in the EV market, with around 3.2 million EVs sold. This record increased by 2 million EV units compared to 2020, more than the combined increase of all other regions.

This growth resulted in China regaining its dominance in the global EV market with an overall share of 53%. There is a massive jump in China's EV penetration rate from 6% in 2020 to 13% in 2021. This growth reflects the government's role, the adaptability of consumers, and a diverse product offering from OEMs.

Analysts view the rise of China's domestic auto industry as a forerunner to a tectonic shift in the global auto market as Chinese EV makers start to sharpen their focus on export markets. In 2021, China, the world's largest car market, exported more than half a million electric vehicles, more than double the figure for 2021.

China's electrifying performance

BYD, the Chinese auto group backed by Warren Buffett's Berkshire Hathaway, has dethroned Elon Musk's Tesla as the world's biggest electric vehicle producer by sales, signaling China's rising dominance of the sector. Shenzhen-based BYD sold 641,000 vehicles in the first six months of 2022, a more than 300 percent jump from the same period a year earlier.

That compared with 564,000 vehicles sold by Tesla, which has blamed a tough second quarter on supply chain and sales disruptions in China after coronavirus lockdowns and travel restrictions hit its operations.

Many of BYD's models are plug-in hybrid vehicles, which use a large battery and a traditional engine for longer journeys. Still, they fall into the category of "zero emission" vehicles under China's sales rules.

According to a report by *CleanTechnica*, sales of electrified vehicles in May 2022 hit 31 percent of the overall market, with 25 percent being fully electric vehicles. In raw numbers, over 403,000 new electrified vehicles hit the Chinese market in May, growing 109 percent compared to the same month in 2021.

The vast majority of the EVs sold were affordable. As expected by anyone who has watched the Chinese market, the top spot was the Wuling HongGuang Mini EV, a cheap and practical EV made by none other than a joint venture with General Motors, SAIC Motors, and Liuzhou Wuling Motors.

With over 600 electric vehicle (EV) start-ups in China, the country is famous as the EV capital of the world. Over 3.3 million EVs were sold in China in 2021, a whopping 154% increase compared to the previous year. Dominating the sales were Shenzen-based BYD and Tesla.

China's current tier-1 EV players in 2022 are Nio, XPeng, Li Auto, Leapmotor, and Nezha, with new players such as pure internet companies with positive capital flows and comprehensive technology reserves entering the industry as tier-2 players.

EV producers in China, which already account for 40 percent of the world market, are in an enviable position to benefit from the strong growth expected in the rest of the Asian market thanks to the signing of regional trade agreements.

As if this were not enough, the production of the heart of EVs—batteries, which represent more than 40 percent of the total cost of an EV—is also concentrated in Asia, South Korea, Japan, and China. Again, China leads in battery components after a long struggle with Japan for component supply dominance. Thus, China's market share for battery components increased from 43 percent in 2014 to 60 percent in 2020, thanks to tight control on the supply of raw materials such as lithium, cobalt, and nickel, for which prices have grown exorbitantly in recent years.

China's growing EV industry is holding up better than that for combustion engine vehicles. It will likely maintain an upward trend in the coming months, despite Covid-19-related lockdown measures and supply chain constraints. According to the China Passenger Car Association (CPCA), in 2022, sales of plug-in electric cars might reach 5.5 million, while the overall New Energy Vehicle volume (including commercial vehicles and buses) should be up to 6 million.

The market share would increase further to 22%, and most plug-ins will likely be all-electric.

China's car exports are taking off, many of them are electric vehicles, and most are going to Europe., China exported half a million electric cars in 2021, and its market share in Europe was second only to Germany's. As the car market goes electric, Europe could quickly find itself running a trade deficit with China in automobiles.

If batteries replace combustion engines and China dominates car production, the disruption will be immense. Automobile manufacturing underpins the prosperity of Europe and Japan. Companies such as Toyota and Volkswagen, plus their supply chains, employ millions of people in stable, skilled manufacturing jobs. They underpin national current account surpluses. A shift in the location of car manufacturing would have an even more significant impact than the past migrations of steel, electronics, or shipbuilding.

More car buyers are favoring EVs as the current high oil pricesoutweigh battery cost increases due to high metals like lithium and nickel prices. This change is especially apparent in the taxi or rental car segment. "Many cab drivers are now reluctant to drive oil-powered cars in the face of escalating oil prices," China Passenger Car Association (CPCA) observes.

Improvements in EV battery performance have also allayed consumer range anxieties: "There has been greater improvement in EV battery performance than in the technology for internal combustion engines," CPCA notes. According to data from China's Ministry of Industry and Information Technology, the average driving range on a single charge has almost doubled from 253 kilometers (157 miles) in 2016 to above 400 km in 2021.

And as the expanding EV market hits critical mass, a more comprehensive array of EV models catering to different budgets, requirements, and tastes is increasing the appeal to Chinese consumers. More foreign automakers like BMW and Volkswagen have joined US EV manufacturer Tesla in rolling out locally made models.

Global carmakers are falling behind Chinese electric car companies in the world's biggest car market, with no foreign brands among the top 10 carmakers in the China EV segment this year, except for Tesla in third place.

EV sales in Mainland China more than doubled in H1 2022 and are now the biggest EV market. 57% of global EV sales were in Mainland China. With 118% year-on-year growth, it is the fastest-growing market as well. China alone witnessed the sale of 2.4 million EVs, 26% of all passenger cars delivered in the country, against just 10% in the first half of 2021.

China's EV Champions

In the world's biggest automobile market, if none of the dominant world brands (except Tesla) are in the top 10 in EV sales, it's something to sit up and notice. Let's look at the leading Chinese brands starting to dominate the EV segment.

BYD

BYD (Build Your Dreams) is the company that sells the most electric vehicles in China. Although most of its sales are among local people in China, it also operates in Europe and America. One of the company's shareholders is Berkshire Hathaway Holding; Warren Buffett is the CEO.

BYD increased its plug-in EV sales by 232% compared to the previous year and broke its record by selling 593,745 rechargeable electric vehicles in 2021. In April 2022, the company announced that it would no longer produce gasoline-powered vehicles and would shift its focus to innovative high-tech batteries, electric cars, and trucks.

BYD's best-selling EV models are Han EV, Song Plus EV, and Yuan EV

SAIC

State-owned and headquartered in Shanghai, SAIC Motor is China's largest automaker. Having brands such as Maxus, MG, Wuling, Roewe, Baojun, Feifan, and IM; SAIC is also part of joint ventures such as SAIC-Volkswagen (Volkswagen, Skoda, Audi) and SAIC-General Motors (Buick, Chevrolet, Cadillac).

Making its weight felt in the sector, especially with HongGuang MINI EV, which is a blockbuster thanks to its affordable price, SAIC is currently the largest EV manufacturer in China. In 2021, it became the world's third-largest plug-in electric vehicle (BEV and PHEV) company, with a market share of 10.5%. The company, which sells electric vehicles under its brands, such as Wuling, Baojun, Maxus, MG, Roewe, and Rising Auto, is also the world's third-largest BEV manufacturer with a 13% global market share.

Changan Automobile

Headquartered in Chongqing, the company's core business is the production of passenger cars, microvans, commercial vans, and light trucks. The company manufactures and sells vehicles under its brands, such as Changan, Oshan, and Kaicene. It has joint ventures with Changan-Ford and Changan-Mazda.

As one of the "Big Four," all state-owned four largest automotive companies in China—Changan sold 2.3 million vehicles in 2021. In 2017, the company announced that it would produce only EVs by 2025 and would end the production of gasoline and diesel vehicles until then. Under its plan to launch 21 electric models in the 2021-2025 period. Changan's main EV line-up consists of two models – Changan Deep Blue C385 and Avatr 11, which it makes in partnership with Huawei and CATL.

Geely

Brands affiliated with Geely broke the company's record by selling over 2.2 million vehicles worldwide in 2021. Thus, Geely became the top-selling Chinese automobile manufacturer of the year across the globe for the fifth time in a row.

There are many EV models of the brands operating within Geely, and EV sales increased by more than 70% in 2021 compared to the previous year. The latest EVs from the Geely stable are Volvo XC40 Recharge P6, Geometry EX3, and Lotus Eletre.

XPENG Motors

XPENG is famous for producing EVs similar to some Tesla models and is known as "Chinese Tesla." The company, which went public in 2020, also draws attention with its semi-autonomous driving system called XPILOT. In 2021, XPENG introduced the prototype of the flying electric vehicle, which it plans to launch in 2024.

XPENG sold 98,155 electric vehicles worldwide in 2021 and increased its sales by 263% compared to the previous year.

The noteworthy models of XPENG are P5 600P, G3 520i, and P7 RWD Super Long Range

NIO

The company defines itself as a pioneer and leader in the premium intelligent EV market. Apart from China, it operates in the USA, Germany, England, and Norway.

NIO focuses on autonomous driving, digital technologies, electric powertrains, and new-generation battery technologies. In 2021, the company increased its sales by 109% compared to the previous year by selling 91,429 electric vehicles worldwide.

NIO draws attention, especially with its fully automatic "Battery Exchange Stations." These stations replace the discharged battery with a full battery and propose the charging problem of EVs. While the company has over 800 battery exchange stations in China, the first station in Europe opened in Norway in January 2022.

Leapmotor

Leapmotor conducted intensive R&D activities in its first four years to develop its own electric/intelligent vehicle technologies. By 2020, it introduced the first artificial intelligence chip of a Chinese car brand designed for use in autonomous driving. Leapmotor is among the pioneers of the intelligent automobile industry, with more than 1200 patent applications.

Leapmotor manufactures at the 'Jinhua AI" factory in China's Zhejiang province. By selling more than 30.000 EVs in the first four months of 2022, the company became one of the major players in the Chinese market in a short period. The company, which has overgrown thanks to its technology development, will soon launch the first vehicle with a Tesla-style CTC (Cell-to-Chassis) battery. This technology will enable Leapmotor to produce lighter cars with a more comprehensive interior than their competitors, and the battery's energy density will increase significantly.

Hozon Auto

In 2018, Hozon Auto obtained a production license from the Ministry of Industry and Information Technologies of China. The same year, it officially introduced the electric vehicle brand NETA and started production with the NETA N01 model.

In 2021, NETA sold 69,174 electric vehicles in China. It increased its sales by 360% compared to the previous year. The company has more than 700 patent applications filed to date.

Besides these major players, many start-ups are mushrooming in China. Some interesting newbies are Airways, Skywell, and Arcfox.

The homegrown Chinese companies are stiffly competing with western automakers in the EV market in China. Their product offering is unique and differentiated. Supported by internet giants, these start-ups have equipped their EV models with the state of the art EV intelligence systems, converting cars into smart devices on wheels. Some of these start-ups employ more tech gurus than car engineers. They have adopted avant-garde sales & marketing strategies and ditched the dealership model to connect with their customer base to create a community of consumers who promote their products by word of mouth. The EV space is becoming crowded with various choices, benefiting consumers the most.

A differentiating factor in the China automobile market is that the Chinese EV consumer profile and behavior are notable and distinct from other parts of the world. Some unique differentiators are :

- Many Chinese customers can quickly adapt to EVs because they have no prior car ownership experience, so there is no 'switching cost' involved. In contrast, consumers in western countries have years of experience driving internal combustion engine cars, making them slow to change their behavior.

- Chinese consumers are price-sensitive, and OEMs have responded by offering EVs at various price points.
- These consumers are keen to adopt the latest digital architecture and intelligent functions that constitute an EV.
- Many enthusiastic Gen Z and millennial buyers take their latest acquisition to another level by personalizing their EVs as a mark of self-expression.

For example, the trendy and bestselling EV model, Wuling HongGuang MINI EV, is a micro car and lends itself to modern after sales transformation. The cult status of Wuling MINI in China shows that there is more than one way to move towards electrification and that automakers need to take the pulse of their consumers.

Revolution on two wheels

China accounted for 53% of global two-wheeler sales in 2021, with electric comprising 70% of the 39 million two-wheelers sold in the country. Implementing new e-bike standards has been a significant sales driver as e-bike owners have to switch to vehicles that comply with new standards

According to Bloomberg New Energy Finance, China's electric two-wheeler sales will increase to 33 million units in 2025 from 30 million in 2021. This growth is by booming food and last-mile delivery business, city-level enforcement of new e-bike standards, and a shift in demand from public transport to private vehicles in the wake of the pandemic. Nearly all recent sales of two-wheelers will be electric after 2035.

China's electric two-wheeler growth is due to two notable policies from the central government. First, in 1999, the government designated certain electric two-wheelers as bicycles. To be considered a bicycle, an electric two-wheeler was required to have a bicycle-style design with functioning pedals, weigh less than 40kg, and have a maximum speed of 20km/h. This classification meant, among other things, that qualifying electric two-wheelers could travel in the bicycle right-of-way, that they did not require licensing and registration, and that users did not need a driver's license. These advantages made motorized travel available to many people who would otherwise be unable to use it.

Second, many cities severely restricted the ownership and use of gasoline motorcycles in their urban cores. In a nation with busy roads and restrictions on gasoline motorcycles, it is perhaps not surprising that a low-cost vehicle that ran on electricity and could travel on bicycle rights-of-way became popular.

As the technology evolved and demand for larger vehicles grew, electric two-wheeler producers began to ignore vehicle speed and weight limits. The regulation was lax, and consumers wanted faster and heavier two-wheelers miming gasoline scooters. Soon these rapid and heavier two-wheelers were common.

Higher speeds have undeniable benefits to users, but the higher pace can also have costs in the form of safety— both the safety of users and the public. In response to the perception that two-wheelers had become unsafe, several cities imposed more stringent regulations, forbidding electric scooters in urban centers while gradually allowing higher speeds and weights in outlying areas. And in December 2009, the central government reiterated its commitment to the "20/40 rule," mandating that two-wheelers maintain a 20km/hr speed limit and 40kg weight to fall into the category of electric bicycles. Faster and heavier vehicles still exist, but they are now classified as electric motorcycles, requiring registration and driver licensing.

Leading OEMs dominate more than 50% share in China's electric two-wheeler market. Key participants are Yadea, Aima, Tailg, Luyuan, and Xinri. Other notable OEMs include Niu and Ninebot, which have entered the electric two-wheeler market through premium offerings and exports. The personal purchase of electric two-wheelers is mainstream, driven by old product replacement (66.5%) and new user demand (19.6%).

The battle for batteries

As the world transitions to electric vehicles (EVs), companies are racing to secure and strengthen their positions in the battery supply chain, from mineral extraction and processing to battery and EV manufacturing.

The sector has seen a move towards vertical integration – where one company controls several steps along the supply chain. This transition will ensure guaranteed supply.

Chinese manufacturers command the global electric-vehicle battery market. Indeed, about three-quarters of the world's EV batteries are made in China.

China's share of the market for lithium-ion batteries could be as high as 80%, according to estimates from BloombergNEF. Six of the ten biggest EV battery producers are in China—one of them, CATL, makes three out of every ten EV batteries globally. That dominance extends through the supply chain. Chinese companies have signed preferential deals with lithium-rich nations and benefited from massive government investment in the complex steps between mining and manufacturing. That's made the rest of the world nervous, and the United States and Europe are now scrambling to wean themselves off Chinese lithium before it's too late.

An electric car battery has between 30 and 60 kilos of lithium. It's an estimation that by 2034, the US alone will need 500,000 metric tons of unrefined lithium a year for EV production. The stranglehold of China on the lithium supply chain makes Western efforts to expand battery production capacity more imperative than ever.

Those efforts are taking shape, albeit slowly. If everything goes to plan, there will be 13 new Giga factories in the United States by 2025 and an additional 35 in Europe by 2035.

But those Giga factories will need lithium—and lots of it.

Also, there's an important piece missing between mining and manufacturing. Turning lithium ore into the purer lithium carbonate or lithium hydroxide needed for batteries is an expensive and complex operation. It takes years to get a lithium processing plant or Giga factory off the ground, and it could take decades and an estimated $175 billion for the US to catch up to China. China controls at least two-thirds of the world's lithium processing capacity, and this, more than anything, could give it a stranglehold on the battery market for years to come.

Lithium pulled from new mines in the US and Europe might need to be shipped to Asia and back again for refinement before using it in electric cars: increasing emissions, compromising energy independence, and handing China a trump card. All this needs immediate investments.

In the global battery supply chain, China is everywhere. Tianqi Lithium, a Chinese mining and manufacturing company, controls almost half of the world's lithium production. Tianqi Lithium also owns stakes in SQM, Chile's biggest mining company, and Greenbushes, Australia's biggest lithium mine. Both Tianqi Lithium and its domestic rival Ganfeng Lithium have signed deals across South America's "lithium triangle," a mineral-rich part of the Andes at the junction of Argentina, Bolivia, and Chile. It's a similar story for other rare-earth materials needed for batteries: China controls 70% of the mining industry in the Democratic Republic of the Congo, home to almost all of the world's cobalt, another critical component of lithium-ion batteries.

In addition to locking down global lithium supplies, China has also started to expand domestic production; it's now the third biggest producer of lithium behind Australia and Chile, even though it holds less than 10 percent of the world's supply.

The number of Giga factories in the pipeline has surpassed 300 as the global rush to develop lithium-ion battery production capacity continues. There are 6,387.6 gigawatt hours (GWh) of lithium-ion battery capacity in the pipeline, as assessed in Benchmark's May 2022 Gigafactory Assessment, a 68% year-on-year increase.

China continues to dominate Giga factory announcements, but North America and Europe have seen significant growth over the past two years partly to automaker and battery manufacturer joint ventures.

China remains the dominant builder of Giga factories, with 226 due to be operational by the end of the decade, making up over 75% of all facilities tracked in Benchmark's Gigafactory Assessment. In 2031, China will have almost 4,500 GWh of gigafactory capacity, representing 70% of the global total.

14

NORWAY

"You may go where you want, but you can't escape yourself."
—Norwegian Proverb

EV sales in Norway made a record in 2021: Almost 65% of all new cars sold were fully electric. In addition, 22% were plug-in hybrids. Put differently, only 14% of new vehicles sold were without a plug.

2021 turned out to be the year EV sales skyrocketed in Norway, where ground-breaking records became frequent. In December 2021, electric vehicles had a market share of 67%; however, September was the record month of 2021, with a staggering 77.5% electric market share.

After 11 years in a row with Volkswagen residing at the throne of the Norwegian car market, 2021 was the year for a new car brand in the number one spot: Tesla was the most-sold car brand in Norway. The explanation is that Tesla Model 3 was Norway's bestselling car model in 2021. Next in line was the plug-in hybrid Toyota RAV4, followed by Volkswagen ID.4 and Tesla Model Y.

Norway's bustling EV market shows no signs of slowing down. By 2023, all taxis will be emission-free, and the wireless system will help support this goal. On top of this, Norway is on course to reach its central objective to have all new cars sold be zero emission – meaning they are battery electric or hydrogen-powered, by 2025.

Norway shows the way

Norway's leading position in the EV sector required the country to overcome cost and infrastructure challenges inherent to adopting new, disruptive technology. The tax schemes have helped to offset the higher production cost of EVs, making electric cars competitively priced with gas models. On top of this, EVs have a lower lifetime cost of ownership: another advantage over gas cars. The country's adaptability has been a considerable denominator in overcoming such challenges. It has quickly worked to extend its electric infrastructure. It continues to develop new technologies such as wireless charging and new batteries, hoping to lower costs and phase out fuel cars.

Oslo has been an EV testing hub for the rest of Norway, and Norway itself has been a success story for the rest of Europe and the world to follow. The cold and rural climate proves that EVs and their batteries can thrive in less agreeable weather. By starting early with incentives and adapting infrastructure and technology to fit demand, Norway has fostered an eco-conscious society that sees EVs as the easy and natural choice.

Norway made a new all-time monthly record of all-electric car registrations with 13,983 new BEVs in March 2022. (up 62%), which translated into a record share of 86.1%. No other country can match Norway in terms of all-electric car share.

At the beginning of the 1990s, Norwegian politicians made the most progressive EV policy in the world by introducing purchasing incentives for EVs – long before there was a mass EV market. Norwegian politicians have held a steady course, consistently finding new policy measures, regulations, and support schemes.

Norway's EV history is one of the most significant climate successes, thanks to consumers embracing the electric transition and purchasing zero-emission vehicles.

So, how did Norway become the world's top-selling electric-vehicle market per capita? Not because of suitable conditions and not because Norwegians are more environmentally friendly or concerned about climate change. We can credit strong demand-side policies kept in place for a long time. Therefore, the government taxes the sales of new polluting cars heavily but does not tax EVs at all, making EVs more expensive because of their production costs, a competitive and appealing option.

A slew of incentives by the Norwegian government to promote EV adoption over the past 28 years has finally borne fruit. The zero-emissions incentives include:

- No purchase/import taxes (1990)
- Exemption from 25% VAT on purchase (2001)
- Low annual road tax (1996)
- No charges on toll roads or ferries (1997 and 2009)

- Free municipal parking (1999)
- Access to bus lanes (2005)
- 50% reduced company car tax (2000)
- Exemption from 25% VAT on leasing (2015)

The country, joined by Europe's No. 2 in electro-mobility, the Netherlands, intends to phase out all fossil fuel-powered automobiles by 2025. Elon Musk, CEO of the US electric car company Tesla Motors, responded to Norway's goal by tweeting: "What an amazingly awesome country. You guys rock!"

Norway illustrates that with incentives that eliminate the price advantage of conventional gas-burning vehicles, many people will go for the electric option. And since 98% of Norway's electricity comes from hydropower, the country's burgeoning EV fleet leaves almost no carbon footprint.

Thanks to its lucrative offshore oil and natural gas business, Norway can afford to promote e-mobility with generous incentives, including the considerable bonus of exemption from a 25% sales tax. Norway's access to abundant and cheap zero-emission hydroelectric power means it can offer e-car owners free power charging at public charging stations.

But the government recently announced a new plan to roll back VAT exemptions, especially for higher-end vehicles, to slash more benefits potentially.

Lessons from Norway

Rapid changes in the Norwegian car market provide a glimmer of hope for a quick energy transition to net zero. The net-zero pathway from the International Energy Agency (IEA), published in May 2021, suggests that by 2030, 60% of new car sales should be BEVs – a milestone Norway has already passed.

Norway took ten years to move from 1% to 65% market share for EVs, and it is all set to pass an 80% share of new vehicle sales within the next year!

The Norwegian experience shows that an incentive scheme composed of both push and pull factors is a must to succeed in this field. The EV-friendly car taxing system and the possibility to drive EVs cheaply (i.e., escaping road fees and using bus lanes) have enabled the massive uptake of EVs in Norway. The country's ability to ensure the further building of fast charging stations has been key to building consumer confidence and continuing EV success.

One may argue that Norway, with a population of 5.3 million people, produces 95% of its electricity needs by hydropower. It is not an example of EV adoption which can be replicated in other parts of the world, especially in large countries like the USA, China, and India. But this country has important lessons if the world is serious about fighting climate change and securing a clean environment for future generations.

Norway's electric car policies are hard to imitate. Norway can be generous because high oil and gas production revenues have helped it amass the world's biggest sovereign wealth fund, worth $1 trillion.

The Norwegian transportation sector is heavily taxed, which includes registration taxes on new vehicles, annual taxes, taxes on fuels, and numerous toll roads. Because of a wide range of EV incentives, the purchase price matches or is even less than ICE cars–making it a no-brainer for customers to opt for electric vehicles.

It may sound strange to many, but there is no 'master plan' that caused the Norwegian Ev's success! There is no dedicated electric vehicle strategy or program, but rather a series of policy interventions and programs contained in climate policy documents, national transport strategies, and budget plans designed to encourage EV uptake.

The high growth in BEV penetration by Norway's long-running policy commitments has made BEVs less of a novelty for citizens, thereby reducing perceived risk. As the technology matured and improved over time, the burdens of owning a BEV, such as high upfront costs, more significant discomfort, less safety, and a lack of charging infrastructure, were progressively removed. Early 'trial-and-error' experiences became a reference point for citizens now observing dramatic improvements in range, reliability, availability, and cost.

While the reasons for purchasing BEVs may differ among individual purchasers, the key incentives are ultimately practical/economic rather than ethical/ environmental. One of the most prominent analysts of electro-mobility in Norway has concluded that "in general, VAT exemption and access to bus lanes are the two most important incentives for BEV sales in Norway." This decision by the government implies that generous economic incentives can overcome other diffusion barriers–such as anxieties related to vehicle range, charger availability, and overall convenience–at least in the early period of technological uptake.

Even if the Norwegian experience shows that EV owners love their electric cars and use them daily, the technology has not come far enough to be competitive without a comprehensive package of incentives to promote zero-emission vehicles. The consumers are conservative and still need economic and other incentives to choose electric when they buy a new car.

The government set identifying letters, "EL," for use on license plates of electric vehicles to simplify any control of whether the vehicle fulfills the conditions for user incentives such as free parking, free passage through toll booths, etc. In July 2016, the "EL" prefix plates exited, and the first electric vehicles registered with the new "EK" series were on the road.

Norway is now a global test lab for electric cars and the new consumer behaviors that the transition to electric creates. It has been clear for some time that for car producers, moving to EV is not just about regulatory pressure. In Norway, it's increasingly apparent that consumers may prefer electric vehicles once price parity is under control. This control could contribute to a considerable "pull" effect.

15

INDIA

"Be the change that you wish to see in the world."
—Mahatma Gandhi

Not to be left behind in the action of global developments, the government of India has also

formulated a vision for increasing EV penetration. With the primary objective of reducing fuel import bills and CO2 emissions, the government of India has taken specific steps to promote electric mobility adoption.

India's energy import bill ballooned to $119 billion in the fiscal year ended 2022, nearly double that of the previous year. As per the forecast by IEA, this will triple by 2040. Annual gas imports would cost $43 billion, while oil imports would cost $250 billion in 2040.

The shift to EVs will help reduce India's energy imports as it looks to cut the oil import bill by half by 2030 and reduce emissions as a part of its commitment to the Paris climate treaty.

The Indian automotive industry is the fifth largest in the world and will be the third largest by 2030. Catering to a vast domestic market, reliance on conventional modes of fuel-intensive mobility will not be sustainable. India benefits on many fronts by shifting towards electric vehicles (EVs): it has abundant renewable energy resources and an available skilled workforce in the technology and manufacturing sectors.

The Government of India has put in place several fiscal and non-fiscal measures to facilitate the adoption of electric mobility.

NEMMP

The National Electric Mobility Mission Plan 2020 (NEMMP) was launched in 2013 by the Department of Heavy Industry (DHI) as a roadmap for the faster manufacture and adoption of EVs in India.

NEMMP is a culmination of comprehensive, collaborative planning for promoting hybrid and electric mobility in India through policies aimed at gradually ensuring a vehicle population of about 6-7 million electric/ hybrid vehicles in India by 2020. It is a composite scheme using different policy levers such as:

1. Demanding side incentives to facilitate the acquisition of hybrid/electric vehicles
2. Promoting R&D in technology, including battery technology, power electronics, motors, systems integration, battery management system, testing

infrastructure, and ensuring industry participation in the same

3. Promoting charging infrastructure
4. Supply-side incentives
5. Encouraging retro-fitting of on-road vehicles with the hybrid kit

FAME

In 2015, the government launched Faster Adoption and Manufacturing Hybrid and Electric Vehicles (FAME). This scheme offered incentives for clean fuel technology cars with the long-term objective of boosting their sales.

However, despite incentives as high as INR 140,000 on some cars, the scheme received a lukewarm response. Sales of electric and hybrid cars contributed to only a fraction of the 3 million passenger vehicles sold in India in 2016. According to the Society of Manufacturers of Electric Vehicles, EV sales in India stood at 22,000 units for the year ending 31 March 2016, of which only 2,000 units were four-wheelers. This figure climbed to 25,000 units for the year ending 31 March 2017, with four-wheelers contributing a mere 8%. Though the EV industry witnessed a growth of 124% selling 56,000 units in the fiscal year 2017-18, electric cars slumped by 40% to 1,200 units.

In March 2019, the Government released the Phase-II of the FAME India Scheme. To be implemented over three years with effect from April 1, 2019, the FAME-II scheme has a total outlay of Rs. 10,000 crores across three verticals :

1. Demand Incentives
2. Establishment of a network of charging stations
3. Administration of the Scheme, including publicity activities

On June 11, 2021, the Ministry of Heavy Industry announced further amendments to the FAME II scheme to boost EV demand among consumers.

The Department of Heavy Industries has also sanctioned 2636 charging stations in 62 cities across 24 States/UTs under the FAME India scheme phase II.

The incentives for various vehicle categories under FAME II are as follows:

S.No	Total Approximate Incentives	Approximate Size of Battery
1	**Two Wheeler:** Rs 15000/- per kWh up to 40% of the cost of vehicles	**Two Wheeler:** 2 kWh
2	**Three Wheeler:** Rs 10000/- per kWh	**Three Wheeler:** 5 kWh
3	**Four Wheeler:** Rs 10000/- per kWh	**Four Wheeler:** 15 kWh
4	**E Buses:** Rs 20000/- per kWh	**E Buses:** 250 kWh
5	E Trucks: Rs 20000/- per kWh	

Besides the Central Government, over 27 states and UTs have formulated strategy plans for transforming mobility to provide their citizens with safe, inclusive, economic, and clean transportation options. While some states like Karnataka and Tamil Nadu have had a head start due to pre-planned public policies, targeted investor incentives, and support infrastructure, other states have drafted policies to stimulate market demand and create infrastructure.

India rolled out the website e-AMRIT – https://www.e-amrit.niti.gov.in/ – at the COP26 summit in Glasgow in November 2021, which functions as a one-stop destination for all information on electric vehicles. It addresses key concerns about the adoption of EVs and their purchase – such as charging facility locations and EV financing options, information about investment opportunities, government policies, and available subsidies for drivers and manufacturers.

The story so far

Although the electric vehicle market in India is in its nascent stages at present, it is poised to emerge as one of the leading electric vehicle markets in the world.

The two-wheelers (scooters, motorbikes) and three-wheelers (autos and rickshaws) mainly run the automotive sector in India. They play a significant role in last-mile mobility in the country.

A record 498,000 EVs were sold in India in FY2022, per the Society of Manufacturers of Electric Vehicles (SMEV). Of this, 345,000 were electric two-wheelers, and 133,000 were electric three-wheeler.

But in the April-August period of FY2023, EV sales crossed the 400,000 milestone – which is 80% of the entire FY2022! At this rate, EV sales could leapfrog toward the 750,000 total sales mark in FY2023.

In the electric 2-wheeler market, Hero Electric is the market leader, and Okinawa Autotech is closely following Hero Electric. Ampere Vehicles, TVS Motors, Bajaj Auto, and Ola Electric's newbie other electric vehicle companies. There was a sale of more than 44,000 high-speed electric 2-wheelers in India. IT was just over 13,000 units in July 2021, a massive jump of 230% year-on-year! And the share of electric 2-wheelers reached 4.26% of the total 2-wheeler sales in July 2022.

YC electric, Mahinda & Mahindra, and Saera electric dominate the electric 3-wheeler segment. In July 2022, the sale of electric 3-wheeler passenger vehicles crossed 26,700 units, registering a growth of over 168% over the same month of the previous year. Additionally, consumers purchased some 2,700 electric 3-wheeler cargo vehicles

Regarding electric 4-wheelers, Tata Motors is the undisputed leader with its Nexon & Tigor EV models. In July 2022, Tata Motors sold over 4,000 EVs, garnering a near-monopolistic 90% market share. MG Motor is a distant second, with a sale of 257 units in July 2022. Hyundai, BYD, and Mahindra are selling minuscule volumes as of now.

3-wheelers lead the electrification of vehicles in India, and of late, by 2-wheelers. Electric cars are in a very nascent stage. Unfortunately, the most significant player in the car market, Maruti Suzuki, is conspicuous by its absence in the electric car segment. However, the reservation on the part of Maruti Suzuki largely stems from its Japanese partner's global strategy of focusing on hybrid electric vehicles (HEVs) rather than pure battery-operated

EVs. Despite this, Maruti Suzuki has committed to investing US $1.3 billion in India's EV space.

Where do we go from here?

India currently has a low EV adoption rate of about 2%, mainly due to the significant price differential with traditional engine counterparts and the absence of an adequate EV infrastructure. Will India be able to shape its path to e-mobility or merely remain a slow adopter of solutions developed abroad?

In August 2022, Arthur D. Little, the reputed management consulting firm, published a report on India's EV journey until the date and made forecasts of the way ahead.

As per their report, Arthur D. Little (ADL) believes that India's EV industry will cross sales of 10 million units by 2030, with an overall adoption rate of more than 30% across different vehicle classes. Furthermore, more than a third of all two-wheelers sold by 2030 will be electric. However, EV adoption for passenger cars is likely to be a mere 10% by that time, amounting to just 5% of total EV sales.

A report by ACMA-McKinsey released in September 2022 stated that about 10-15% of cars will be electric by 2030. The study estimates that new electric three-wheelers and two-wheelers could grow to 50% and 70% of the total sales by 2030.

Multiple factors are behind India's low EV adoption rates. Fundamental pressures include higher up-front EV prices compared to ICE vehicles, the absence of charging infrastructure (e.g., charging stations, battery swapping centers, chargers, and the entire EV supporting ecosystem), a lack of vehicle options, and low consumer confidence.

India had 592 start-ups functioning in the EV industry as of March 2022, according to the investment research firm Tracxn, working in the domains of battery manufacturing, charging infrastructure, and battery recycling, to name a few.

EV adoption will likely create 10 million new jobs in India by 2030. Consequently, there will be increased demand for upskilling and training the existing workforce. In addition to employment generation, EV adoption will enormously impact India's GDP. According to a study, the contribution of EVs to India's GDP will surpass US$150 billion by 2030. The growth in ancillary industries, such as battery and component manufacturing and IT services, will complement the development.

The increasing use of EVs in India will raise demand for Li-ion batteries, boosting the manufacturing industry. To attain more than 30% EV adoption, as predicted, India will require approximately 800 GWh of batteries by 2030. To meet this rising demand, India is accelerating plans to manufacture Li-ion cells within the country, anticipating US$2.3 billion in government subsidies and more than US$7.5 billion in investment potential, according to ADL analysis. For India to become a global EV manufacturing powerhouse, it will require more than US$10 billion of focused investments in this industry.

Besides the benefits of a cleaner environment, more investments, jobs, innovations, and health benefits, ADL estimates that with a target of 30% penetration of EVs, India's import bill will reduce by almost US$14 billion in 2030.

Plans of Indian Automakers

Tata Motors, the current leader in electric cars, plans to launch 10 EV models by March 2026, investing about $2 billion, of which $1 billion has come from private equity group TPG.

It plans to launch electric cars with a minimum range of 500 kilometers (310 miles) and advanced technology features, targeting buyers in India and beyond. The first electric vehicles (EVs) built using its new platform should be on the road in 2025.

Mahindra & Mahindra in August 2022 unveiled its new electric vehicle architecture called 'INGLO EV platform' and five e-SUVs under the two EV brands - Iconic brand XUV with the Twin Peak logo and the all-new electric-only brand called BE - showcasing its vision for the future of electric mobility.

The company also showcased five e-SUVs under the new brand: the XUV.e8, XUV.e9, BE.05, BE.07, and BE.09. Mahindra also revealed the launch dates for its new e-SUVs and said the first to arrive would be the XUV.e8 in December 2024, followed by the XUV.e9 in April 2025. On the other hand, its BE-branded vehicle BE.05, will debut in October 2025, while the BE.07 in October 2026. The company hasn't revealed any launch date for the BE.09.

This unveiling came a month after the auto giant raised ₹1,925 crores from British International Investment, the UK's development finance institution, for its newly incorporated EV subsidiary. The company plans to invest $1 billion in its EV business between FY24 and FY27.

In early September 2022, Mahindra unveiled its all-electric new model, the XUV 400, and it will be ready for delivery in January

2023. The XUV400 is a direct competition to Tata Motor's Nexon EV and comes with a 39.4 kWh battery pack, and it gets its power from a front-axle mounted electric motor that produces 150 hp. Mahindra claims a range of 456 km on a single charge as per the Indian driving cycle.

Maruti Suzuki wants to launch its first electric vehicle in 2025, which will come from Suzuki's Gujarat factory. Its parent, Suzuki Motor Corporation, plans to invest around 150 billion yen (approximately INR 10,445 crore) by 2026 to manufacture battery electric vehicles and EV batteries in Gujarat. Suzuki has already developed a Lithium-Ion battery pack production unit in a joint venture with Japanese companies Denso Corp and Toshiba

Maruti Suzuki, which had been taking a back seat to EVs, and betting more on hybrids, has now committed itself to pure electrics. That's good news for the Indian mass market.

Hyundai is planning to launch six new battery electric vehicles in India by 2028, for which it has earmarked investment to the tune of INR 4,000 crore. The vehicles will cover a cross-section of segments from mass to premium, hatchbacks to SUVs. Hyundai was one of the first foreign carmakers to launch an EV in India in 2019, bringing the Kona EV under the CKD route. But the market response for Kona was relatively tepid, chiefly due to its high price.

Hyundai is working on a compact, affordable electric car specifically for India. Until its small EV hits the roads, Hyundai will launch premium models, starting with its Ioniq 5 electric crossover in 2022 and gradually moving down the price chain.

Kia has recently launched its first EV, called EV6, in India. The car is being imported to the country as a CBU in limited numbers. But the company also plans to launch an India-focused EV by 2025.

MG Motors has already introduced its ZS EV model in India and plans to bring another electric car in early 2023. The model will be the MG E230 – an affordable, two-door EV – aimed at the mass-market segment. MG Motor plans to get 25% of its total sales from these two EVs in 2023 –an ambitious target.

MG India's electrification plans aren't limited to just the ZS and E230. The company plans a full salvo of 5-6 EVs by 2030. For this, the automaker will need considerable investments. MG is plotting a significant EV offensive, which hinges on a separate EV entity the company is planning to set up. The carmaker is looking for external investment to the tune of US $400 million-500 million (Rs 3,000 crore-3,800 crore) in the new subsidiary to fund the development of a raft of new EVs for the Indian market.

BYD, the world's largest electric vehicle company by volumes, plans to enter the Indian mainstream EV market with its Atto3 electric SUV soon. BYD introduced its E6 MPV to corporate and fleet customers in November 2021. The company's Chennai factory Will assemble the upcoming SUV.his factory can make 800-900 cars a month on a two-shifts and produce up to 10,000 units annually.

Ola Electric has also set its sight on entering the four-wheeler EV segment. The company's founder and CEO, Bhavish Aggarwal, shared a teaser image of an electric hatchback and said the car is about to debut in 2024. He claimed it would range around 500 kilometers on a single charge. Aggarwal also said Ola plans to develop two vehicle platforms and six different cars. After its expansion, the Giga factory in Tamil Nadu will manufacture these cars.

Honey, where's my battery?

As India gears up to electrify its vehicles, battery manufacturing remains the most critical segment in the EV-related component industry. A recent ICRA research estimates that investment in battery cell manufacturing in India will exceed $9 billion by 2030.

ICRA estimates that EV battery demand in India will go up to 15 GWh by 2025 and 60 GWh by 2030. However, this seems to be a gross underestimate. As per a report by India Energy Storage Alliance (IESA), the yearly EV battery demand is about to reach 142 GWh by 2030.

To keep up with the anticipated surge in demand for battery cells and packs, the Government of India has recently signed agreements with three companies for incentives under its Production-Linked Incentive (PLI) Scheme for Advanced Chemistry Cell (ACC) Battery Storage. Reliance New Energy, Ola Electric Mobility, and Rajesh Exports are companies.

Tata group is preparing the blueprint to launch a 'battery company' both in India and abroad to transition to being future-ready. Mahindra & Mahindra is also looking to set up a battery-cell company to power its electrification plan.

Among the conglomerates to have entered the market is Amara Raja Group, which announced plans to invest approximately USD 1 billion in the segment over the next five to seven years. A lithium-ion battery plant by Exide along the lines of Tesla's gigafactory is also under work in Gujarat.

Besides the big names, smaller start-ups, like Bengaluru-based Log9 Materials, are hyper-focused on building cells crucial to EV batteries. Others, like Ion Energy, mainly want to develop battery management systems for the batteries. Among the early entrants in the sector is Gegadyne Energy, founded in 2015, which promises batteries with higher energy density and longer life than the customarily used lithium-ion ones.

16

THE OTHER ELECTRICS

"The only way to make sense of change is to plunge into it, move with it, and join the dance."
- Alan W. Watts

The glitz and glamour of electric cars overshadow a seismic change elsewhere, leading to automobile electrification. I'm talking of electric 2-wheelers, 3-wheelers, and buses.

In China, more than 685,000 electric buses and 195 million electric 2-wheelers are on the road. Almost 40% of India's 3-wheeler fleet is already electric. Globally, 44% of new bus sales and 18% of the bus fleet is electric.

As mentioned in Chapter 12, electric two/three-wheelers will continue to be the largest EV fleet among all road transport modes. Asia, where two/three-wheelers are prevalent today, is the primary growth center.

In its EV Outlook 2022, BNEF estimates that the global e-bus fleet will reach 1.75 million by 2040, representing 62% of all buses on the road. Buses and two-wheelers/three-wheelers will achieve the highest EV adoption rates by 2040.

This growth is heartening, especially for countries like India and China, where most of the population relies on micro-mobility and public transportation.

Electric 2-wheelers

Growth in electric two-wheeler sales has witnessed a substantial escalation globally over the past few years — especially in China, thelargest electric two-wheeler market. Vietnam takes the next spot, with electric two-wheelers having 10% of total two-wheeler sales in the country. India could become the second-largest electric two-wheeler market, surpassing Vietnam by 2026, given the country's ambitious plans for the market.

I have already touched upon the electric two-wheeler market in China in Chapter 13. Let's take a look at what's happening in India.

India

The total sales of electric two-wheelers in India, including high-speed and low-speed, in 2021 increased by 132% over the corresponding year, 2020. The industry registered sales of 2,33,971 units against 1,00,736 units sold in 2020.

The high-speed electric two-wheelers, which have speeds greater than 25 km/h and require a full license, registered a whopping

425% growth, while the low-speed ones (less than 25 km/h, no ownership, no registration) grew only by 24%.

Hero Electric, Okinawa, Ather Energy, and Revolt are current market leaders. Ola Electric had a big bang launch of its electric scooter in 2021, and many more start-ups are joining the fray. Simple energy surprised everyone by announcing it would build a 12.5 million unit electric 2-wheeler factory in Tamil Nadu.

The expected cumulative capacity for electric two-wheelers by 2026 will exceed 30 million units a year! It is a staggering number, almost 50% more than the peak sale of 21.2 million two-wheelers in fiscal 2018-19.

According to the consulting firm McKinsey, the Indian electric 2-wheeler market will hit 4.5 to 5 million units by 2025, accounting for 25%-30% of the total market, and 9 million by 2030.

Based on McKinsey's forecast, there will be overcapacity in the electric 2-wheeler market. But in the world of crystal ball gazing, it is anybody's guess as to where the market will end up.

The Government of India wants sales of electric vehicles (EV) to reach 30% market share for private cars, 70% for commercial vehicles, and 80% for two-wheelers by 2030. Suppose the overall electric 2-wheeler market grows to 40 million units by 2030. In that case, we are talking of 32 million electric 2-wheelers –the current expected production capacity will be then.

The government of India's think-tank Niti Aayog has forecast a 100% penetration of electric two-wheelers (e-2W) in the Indian market by FY 2026–27 in case of an 'optimistic scenario.'

Under this scenario, all factors that positively influence the adoption of e-2W have to be present. The crucial demand incentive is assumed to be in place throughout the period, and the battery cost is to reduce by 8% annually. The report noted that the vehicle's range and power have to improve by 5% from FY 2024 to FY 2026 and by 10% in FY 2027. Niti Aayog forecast that e-2W sales might cross 22 million units in FY 2028–29 under the 'optimistic,' 'same performance,' and 'battery cost challenged' scenarios.

Electric 3-wheelers

An electric three-wheeler, an e-rickshaw, operates on high torque electric motors and longer durability batteries. It offers last-mile connectivity, easy access to narrow roads, and a noise-free travel experience. It helps improve urban air quality by reducing greenhouse gas (GHG) and other emissions. Besides this, as it also aids in lowering additional costs associated with fuel consumption, the electric three-wheeler is gaining immense traction as a convenient and eco-friendly mode of public transportation across the globe.

Factors include an inclination toward the use of electric three-wheeler as an eco-friendly & efficient solution for commute, a rise in the trend of last-mile connectivity, growth in the direction of shared mobility, and an increase in demand for affordable commercial vehicles supplementing the development of the global electric 3-wheeler market.

The electric three-wheeler market has three components power, vehicle, battery, and region. The global market has three types 1000 W, 1000 W to 1500 W, and above 1500 W. All categorizations depend on the power type. The global market has been segmented based on vehicle type: passenger carrier and load carrier. Based on battery type, the market across the countries is of three types lithium-ion and lead acid. The worldwide market across North America, Europe, Asia-Pacific, and LAMEA provides an exciting ground for research and studies.

The key players profiled in the global electric three-wheeler market include Atul Auto Limited., Chongqing Zongshen Tricycle Manufacture Co. Ltd., J.S. Auto Pvt. Ltd., Kinetic Green Energy & Power Solutions Ltd., Mahindra & Mahindra Ltd., Piaggio & C.SpA., Scooters India Limited, Terra Motors Corporation, Bodo Vehicle Group Co. Ltd., and E-Tuk Factory BV.

The global electric three-wheeler market anticipates reaching USD 831.10 million in 2028 from USD 582.46 million in 2021 at a Compound Annual Growth Rate (CAGR) of 5.2%. The Asia Pacific is likely to remain at the forefront and hold the most prominent position in the market during the forecast owing to the high demographic growth, a larger middle, and lower-middle-class income population, and increasing inclination towards electrification vehicles in this region.

India

In FY 2021-22, India sold more than 3,84,000 units of 3-wheelers, including cargo and passenger categories. Nearly half of this – 47% were electric, notching up a volume of more than 1,79,000 units. There was a growth of 94% compared to the previous year.

It's important to note that of the total electric 3-wheelers, the low-speed (L3) category contributed to 86% of the sale. These are known as 'e-rickshaws' and have become ubiquitous, particularly in urban areas.

It's the 3-wheelers that are spearheading India's transition towards electric vehicles. It is an estimation that nearly 90% of 3-wheelers sold will be electric by 2025.

As opposed to its petrol counterpart, which has a running cost of approximately Rs. 4 per km, the electric alternative is around Rs. 0.5 per km. In a country like India, economics drives the last-mile commercial transportation, which is why this significant difference in cost makes all the difference. You cannot beat an electric three-wheeler total cost of ownership game, even with an expensive battery pack.

Electric Buses

According to a report by Markets and Markets, the global electric bus market size has an estimation to be 112,000 units in 2022. It will likely reach 671,000 units by 2027 and grow at a CAGR of 43.1% over the forecast period (2022 – 2027).

The market in the Asia Pacific will experience the fastest growth owing to the high demand for electric buses from countries such as China, Japan, South Korea, and India.

Chinese companies such as BYD, Yutong, Ankai, and Zhongtong dominate the Chinese and global electric bus markets. These players have developed and provided a wide range of electric bus models at lower prices due to the availability of parts and components at cheaper rates. The Chinese government has invested heavily in changing the conventional public transport fleet to electric. As a result, the country dominates the market, ahead of European and American countries.

In its CleanVehicle Directive, the EU has set a target of at least 45 percent of newly registered city buses being equipped with alternative powertrains by the end of 2025 – and at least 65 percent by 2030. However, this figure includes buses with electric powertrains or fuel cells and those running on biofuels or liquefied gas. According to various estimates, 3,500 to 4,000 e-buses (including hybrid buses) are now on Europe-wide roads.

In Europe, more than a dozen manufacturers produce EVs. Some companies are Volvo and Daimler Buses (since 2019), with the eCitaro from Mercedes-Benz, the Polish company Solaris and the mid-sized German firms Sileo and Eurabus. Alongside MAN and Scania, Traton – a VW subsidiary for commercial vehicles – is trying to break into the market.

According to another study released in 2021 by financial consulting company ING, a third of the 200,000 buses in European public transport will be zero-emission by 2030. Zero-emission buses will cover two-thirds of the new city bus registrations in that year.

India is a market with immense potential when even a tiny part of the orders will be electric. By 2025, the research company Interact Analysis forecasts that India will account for more than 10% of the total annual demand for electric buses globally, which is more than Europe and North America combined.

The electrification of buses in India has gained good traction recently.

Convergence Energy Services Ltd. (CESL), a state-owned company, is considering a $10 billion procurement for 50,000 electric buses. Switch Mobility, the UK subsidiary of Indian commercial vehicle manufacturer Ashok Leyland, and Indian transport company Chalo have entered into a strategic cooperation to deploy 5,000 electric buses across India.

In May 2022, Olectra Greentech, a partnership with BYD, said it had bagged the biggest-ever order worth Rs 3,675 crore for supplying 2,100 electric buses from Brihanmumbai Electric Supply & Transport (BEST). Tata Motors has received two significant orders for electric buses in India, following an earlier purchase from the Delhi Transport Corporation. The West Bengal Transport Corporation and the Bengaluru Metropolitan Transport Corporation expect 2,000 electric buses.

Electric Trucks

There was an increase of over 70% in Electric light commercial vehicle (LCV) sales worldwide in 2021. China accounted for the most significant number of LCVs sales in 2021, at 86 000, showing rapid growth consecutively for 2020 and 2021. Europe was the second-largest market, with more than 60 000 electric LCV sales in 2021, where sales have been increasing partially in response to CO2 performance standards.

Korea has seen a very rapid increase in sales of electric LCVs, reaching 28 000 in 2021 (a 12% market share of overall LCVs), up from just 1 500 just two years prior. This increase is attributable to an innovative policy in Korea that incentivizes the adoption of EVs for commercial use.

Most electric LCVs are BEVs; only a few PHEVs are available for sale in Europe. This sale may reflect that most electric LCVs are acquired for specific uses within fixed delivery areas and may not need an extended driving range.

Global sales of electric medium- and heavy-duty trucks in 2021 more than doubled over 2020, while total sales volumes remained at roughly the same level as the previous year. Electric medium- and heavy-duty truck sales totalled more than 14 200 in 2021, representing less than 0.3% of the total registrations for medium- and heavy-duty vehicles worldwide.

China accounted for nearly 90% of electric truck registrations in 2021, down from almost 100% in 2017. Sales in the United States and Europe have begun to rise rapidly in the past few years, driven by increased available models in those markets, policy support, and rapidly improving technical viability and economic competitiveness of electric trucks in specific applications.

17

ELECTRIC CAR BUYER'S GUIDE

" Happiness is not in money, but in shopping."
—Marilyn Monroe

"Should I buy a petrol car or a diesel one ?" The most common question asked by rupee-stretcher customers in India when considering a car purchase.

The answer to that is, "It depends." It depends on the economics, that is. Given the substantial retail price difference between petrol and diesel fuels in the country until very recently, the choice was for the diesel option if one's annual car mileage was above a specific figure. Even though the diesel model is invariably higher priced by the OEMs, simple back-of-the-paper calculations and higher fuel efficiency make a choice clear for you.

Enter the electric car.

Now, there is a new elephant in the room.

The choice just got a little more complex. To further add to the confusion, there are several types of electric cars – hybrids, plug-in hybrids, and batteries only.

Help!

Relax. Help is at hand, and here's a brief guide to assist you in making the right decision for your next prized possession.

Is an electric car right for me?

Available electric car models reached 450 in 2021, numerous launched after that, and plenty in the pipeline. It's raining electric cars!

But beyond the commotion, you need to answer the question seriously – is going for an electric car the right decision for you?

Beyond the noble and humanitarian cause of reducing GHG emissions and curtailing your carbon footprint, many other practical factors influence buyers.

It boils down to three basic questions.

- How much mileage do your cover per day on average? (electrics typically have a range of 200-250 km when fully charged, while the fossil-fuel ones have 300-500 km on a full tank)
- Do you have access to a charging point at home? (electrics would require 6-12 hours of recharging while refueling the ICE cars take a few minutes.
- What are the electricity and fuel costs in your region?

(electricity costs are much less than fossil fuels, and this is the most significant factor in comparing the TCO)

Many customers would also consider the near-noiseless operation of an electric car to be a significant plus compared to the inherently noisier ICE cars. Also, don't forget that charging at home means never going to the gas station. Since electric vehicles require no maintenance like oil changes, filter replacements, coolant changes and so on, it means much less hassle of visiting the workshops.

Want more? Install a solar rooftop at home and further reduce your cost of operations, apart from having your car charged with zero-carbon emission technology!

Pros and Cons of Electric Cars

To take a more balanced view, here are some pros and cons of an electric car:

Pros

- ***Quiet and Quick:*** It will just take one ride in an electric car to experience its quietness and smooth running. It will surely make most regular cars seem clunky. What surprises people is the superior acceleration of the electrics. Step on the accelerator, and the wheels immediately get the power supply. Tesla often brags about this feature.
- ***Home recharging:*** All you need to do is to reach for the plug in your parking slot at night, and when you wake up the following day, you are ready to drive another 150-300

km, depending on your model—no more visits to the gas station.

- ***Cheaper to operate:*** Electricity is affordable and has a significant advantage over petroleum. It is now available 24 hours in most cities and towns. The cost per mile to fuel an electric car is approximately one-third to one-fourth the price of petroleum. Maintenance costs are also far less, which means considerable savings in electric operation.
- ***No tailpipe emissions:*** You get better air quality for yourself, your family and your community because of zero emissions. The emission benefits are dramatic in regions with a potent grid mix of renewables–such as hydro, wind, and solar.

Cons

- ***Limited range:*** Range Anxiety! The worry occurs because most affordable electric cars only have about 80 to 100 miles of range and take hours to refuel fully, and EV advocates argue that 100 miles of range are plenty for most driving. As a result, nearly all electric car drivers rarely, if ever, experience range anxiety. Also, the content of EVs is improving incrementally every year with higher capacity batteries.
- ***Long refueling time:*** It takes several hours to fully charge an electric, while ICE cars get refueled in minutes. Although fast DC charges are now available, this is one considerable disadvantage at this time. And if you forget

to plug in your electricity at night–you need to make other plans for your commute the next day.

- ***Higher Cost:*** Electric cars are considerably more expensive than ICE cars, which is a massive deterrent to EV adoption. But with battery prices expected to continue falling, the inflection point of price parity is achievable in a few years.
- ***Lack of Consumer Choice:*** This is a problem, particularly in India, where only a few electric car models are available, and the choice is severely limited to the consumer. But in developed countries like the USA, over100 models are available as options. And in the next 2-3 years, there will be far more models and variants to choose from, so this is just a transient issue.

Before you buy

Buying an electric car can be a little more complex than purchasing a conventional one.

The following points should help you make the right choice :

- **Price:** Electric vehicles are environment-friendly, but they are not pocket-friendly! They are pricey when compared to their conventional counterparts. So, you should keep your budget in mind before buying an electric car.
- **Incentives and Tax benefits:** The motivations and tax benefits offered by the government can reduce the overall cost of electric vehicles. Apart from lower GST on electric cars, the government gives income tax exemptions on car

loans and offers lower registration charges while registering the vehicle.

- **Specifications:** Electric vehicles come with various electric motors and battery packs of different capacities. You need to choose an electric car with specifications matching your requirements.
- **Driving Range:** The driving range is one of the crucial factors when buying an electric vehicle. Don't rely on the manufacturer's specified driving range figures. Instead, look for real-life reviews to get an idea about the driving range of the EV in real-world conditions.
- **Battery Life:** A battery pack is one of the most critical components of an electric vehicle. At the same time, the battery is also one of the most expensive parts of the EV. Replacing a battery pack of an electric vehicle can burn a hole in your pocket. So, always check the battery life before investing in an electric car.
- **Charging Infrastructure:** You cannot find electric vehicle charging stations as frequently as fuel stations, which is one of the biggest challenges for EV owners. Find out about the charging infrastructure in your area before taking a call.
- **Charging Options:** An electric car comes with a wide range of charging options, such as fast charging, standard charging, and slow charging. Fast chargers are expensive to install, and you can hardly find a few of them in your city. Standard and slow charging technique is feasible and easy to install at home.
- **Additional Costs:** You have to spend extra money to get a home charger, charging station installation, maintenance,

and public station charging. Make yourself aware of these costs to avoid being surprised later.

- **Cost of insurance:** The cost of insuring an electric vehicle is on the higher side because EVs are expensive. Also, insurance companies charge a higher premium on EVs than conventional ones. So do consider this factor.
- **Practicality:** It all boils down to your requirements. How much distance do you drive in a day, and what is the driving range of the electric vehicle? It makes sense to own an EV if you reside in a metro city and only need a car to commute to your office. For others, an electric vehicle might not be a practical choice.
- **Brand and Model:** There are only a handful of electric cars in the Indian market. So, the choice should not be a difficult task: Hyundai, Mahindra, Tata Motors, and MG Motor India retail electric cars in the country. The options are few, but they belong to different price brackets. So, depending on your budget and requirement, you can choose the right brand and model.
- **After-sales support:** Once you purchase an electric vehicle, the ownership experience can be good or bad, depending on the after-sales service and support offered by the manufacturer. Nowadays, most EV manufacturers offer several years of warranty on battery packs and install a wall charger at your desired location. The service centers will have trained professionals to service the electric car. So, always opt for a brand offering good after-sales support for a good ownership experience.
- **Electric or Hybrid? :** If buying an electric vehicle is not practical, you can consider buying a hybrid car. This

segment of the car produces fewer emissions than a conventional car. This vehicle uses an internal combustion (IC) engine and an electric motor to run efficiently. The electric motors draw power from the battery packs, and the regenerative braking or the IC engine charges them.

What to look for in an electric car?

An electric vehicle is a complicated machine with advanced technology and a complex electric drivetrain. It is crucial to understand the basic functionality of an electric car before buying one. Here are the necessary specs to keep in mind while selecting an electric vehicle:

• **Driving Range:** An electric vehicle's range refers to the distance it can go on a single full battery charge. In traditional combustion engine vehicles, this number is equivalent to a full gas tank range. Knowing your EV range is essential for mainly geographical reasons. Depending on your commute to work, school, or wherever you need to go, you will need a specific vehicle range to get you there.

To give you an idea, the Mercedes EQS 450+ has a driving range of 640 km, and Tesla Model S can take you 585 km on a single charge, and the BMW iX xDrive 50 can take you as far as 505 kms. Then we have the Volkswagen ID.4 Pro with a driving range of 410 kms, the Hyundai Kona Electric 64 kWh with 395 kms range, the Tesla Model 3 with a 380 kms range, the MG ZS EV long range with 365 kms, and the Tesla Model Y with 345 kms. At the lower end of the driving range, we have the Audi Q4 e-Tron 35 at 285 kms, Nissan Leaf at 235 kms, and Kia e-Soul at 39 kWh at 230 kms.

• **Battery Life and Warranties**: Like any battery-powered electronic device, electric vehicle batteries don't have an unlimited lifespan. As the battery ages, it degrades and thus will hold less charge when "full." High temperatures and overcharging, among other factors, contribute to faster battery degradation over time.

Most electric vehicles have a battery performance warranty, generally between five and eight years of use or around 100,000 miles of driving. The car battery maintains above a certain percentage of its original capacity. Depending on how much you drive and how long you plan to own your vehicle, warranties on battery life and degradation are essential.

For example, the Tata Nexon EV offers an 8 Years/160,000 kms battery warranty. Nissan Leaf has an eight-year/100,000 miles warranty on the battery, and Tesla Model S comes with an eight-year/ 150,000 miles battery warranty. As you can see, the eight years/100,000 miles warranty on the battery is the norm now.

• **Horsepower:** Many car shoppers are familiar with horsepower because it's a term that has described the power of machines, including gas-powered cars and now EVs. The word dates back to the 18th century when James Watt coined the phrase Horsepower, to make machine power relevant to people familiar with muscle power. (It roughly describes the number of pounds of coal a horse can move over a prescribed period.)

An electric car's spec sheet often shows the max power of a vehicle, not only in horsepower but also in kilowatts (kW). Guess what: it means the same thing, even though the numbers are different.

One horsepower equals 746 watts (or 0.746 kilowatts). Automakers simply convert the more mysterious kilowatt term to the more familiar horsepower rating, multiplying the kilowatts by 1.34 (inverse of 0.746). In the case of the Nissan LEAF, take its 110-kilowatt motor, and multiply it by 1.34, to derive a rating of 148 horsepower.

This calculation is an important metric when comparing electric or fossil-fuelled cars and deciding whether you will make do with a slowpoke or want the thrill of a beast.

- **Torque:** The fun of electric cars is mainly derived not from horsepower, but torque, perhaps the most important but least understood metric for car buyers.

Horsepower or kilowatts describe how fast your EV can travel. But torque applies to how quickly that maximum power is delivered.

Imagine that you're in front of the line at a stoplight. The light turns green, and you step down on the accelerator pedal. How quickly do the axles twist? How fast do you either glide or rocket away from the intersection?

The electric powertrain in an EV—compared to an internal combustion car—is highly efficient at delivering immediate bursts of power to the wheels. In other words, EVs have loads of low-end torque.

Torque, measured in Newton-meters (Nm), is arguably a more critical spec because we commuters regularly feel a vehicle's torque as we zoom around a city. We frequently capitalize on torque, but we rarely test the horsepower capabilities of our cars.

Here's the main point: a higher torque number will mean brisker acceleration, especially from relatively low speeds.

In addition, it's essential to consider that heavier vehicles require more work to move down the road, so two cars with the same amount of horsepower or torque might feel very different on the road. The lighter one is likely to provide a much zippier drive. In some cases, the sky-high torque of an EV will be more than a lightweight car can manage in terms of handling and maneuverability.

• **Battery Capacity (kWh)**: In an electric car, the amount of work you can derive from the battery pack is by the number of kilowatt-hours (kWh) stored in your EV's battery pack.

It's not dissimilar to the number of liters of liquid fuel stored in a tank of an ICE car. If your vehicle has a 40-liter tank and manages 12 kms per liter, a full tank holds enough energy for 480 kms of driving range. Simple enough, right?

The same thing, in rough terms, is happening with your EV's battery. The base version Tesla Model 3 typically goes a distance of 6.5 km per kWh. So, with a 54-kilowatt-hour battery pack, when driven efficiently, it can travel 350 kms on a single charge.

Unfortunately, most of us don't drive all that efficiently, so it's more likely that we'll get around 300 kms on those same 54 kilowatt-hours of energy. Also, very hot or cold weather can reduce the efficiency of the stored energy.

Battery size matters when considering the range of electric cars

- **Charger Rating**: Given the relatively long time it takes to recharge an EV versus how long it takes to pump gas into an internal combustion car, it's essential to give some passing thought to the kilowatt charging rate of an EV.

If your EV has a 6.6-kilowatt charger and you have it plugged in for one hour, your battery will probably get an additional 6.6 kilowatt-hours of energy. Alternatively, if your EV has a faster 10-kW charger, you'll have 10 more kilowatt-hours of juice in that same hour. This charging experience shows the main difference between adding about 20 to 25 miles of range in an hour of charging and about 35 to 40 miles simultaneously.

18

WORLD'S BESTSELLING ELECTRIC CARS

Global passenger plug-in electric car sales amazed in 2021, not only because of the tremendous expansion but also because of expansion in a time when the overall market was struggling.

In 2021, almost 6.5 million new passenger plug-in electric cars got registered globally, which is about 108% more than the 3.1 million in 2020. The average market share improved to about 9% (including 6.1% BEVs), compared to 4% in 2020 (including 2.8% BEVs).

So, almost 1 in 10 cars sold worldwide in 2021 were electric! Furthermore, BEVs had a 6.1% market share with global sales of 4.6 million units in 2021 (up 115% over 2020). PHEVs notched up sales of 1.9 million units in 2021 (up 93% over 2020) with a 2.9% share of total car sales.

The first six months of 2022 witnessed the continued acceleration of electric cars. According to *EV-Volumes* data, more than 4.1

million plug-in cars got registered during the first half of 2022 (including over 913,000 in June). A year ago, at that point, it was around 2.5 million. That's a 65% jump through the roof! The share of electric cars in global sales touched 12% in the first half of 2022.

There are probably more than 500 options for electric cars. The availability of models varies from country to country, and customers have more choices in developed countries.

Let's look at the world's top-selling electric car brands in 2021 and 2022. Here they are, with sales figures rounded off :

Brand	2021 Global Sales	H1 2022 Global Sales
Tesla	936,000	565,000
BYD	594,000	641,000
SAIC-GM- Wuling	456,000	233,000
Volkswagen	320,000	158,000
BMW	276,000	157,000
Mercedes-Benz	228,000	126,000
SAIC	227,000	124,500
Volvo-Geely	189,000	231,000
Audi	171,000	84,000
Hyundai	159,000	106,500
Kia	158,000	113,500

Source: insideevs.com, cleantechnica.com, EV-volumes.com

Tesla was the No.1 EV brand in 2021. But BYD has just overtaken Tesla in the first six months of 2022. The caveat is that the BYD portfolio includes PHEVs, whereas Tesla is pure electric. The SAIC-GM-Wuling triumvirate held on to its No.3 position, but barely so, with Volvo-Geely combine breathing down its neck. VW and BMW are now neck-to-neck in the race, with Mercedes-Benz just behind them.

Now let's look at the top selling models globally in 2021 and in the first half of 2022. These were as below, again with sales figures rounded off :

Model	2021 Global Sales	H1 2022 Global Sales
Tesla Model 3	500,000	219,000
Wuling's Hong Guang MINI EV	424,000	208,000
Tesla Model Y	411,000	315,000
Volkswagen ID.4	122,000	64,000
BYD QIN Plus	112,000	132,000
Li Xiang One EREV	90,000	60,000
BYD Han	87,000	97,000
BYD Song Plus	79,000	159,000

Source: insideevs.com, cleantechnica.com, EV-volumes.com

The significant change in 2022 over 2021 is that Tesla's Model Y has overtaken its sibling, Model 3, in global sales. Wuling's MINI EV is close behind Model 3, and BYD's Song Plus has risen to the No. 4 position. BYD's Qin Plus is the fifth-largest-selling model in the world.

Electric car specifications

Here are the brief specifications of the latest bestselling electric cars :

Tesla Model Y

Tesla Model 3

Tesla Model S

Tesla Model X

Tesla Model Y Long Range: The body styling of Model Y is that of a compact SUV with 5/7 seats. Fitted with a 75 kWh battery, it has an EPA-tested range of 330 miles. The two independent electric motors with a combined output of 480 hp drive the car and digitally control torque to the front and rear wheels.

Tesla Model 3 Long Range: The luxury compact sedan has dual electric motors with a 476 hp power output. The battery capacity is 82 kWh, giving it a driving range of 358 miles.

Tesla Model S: Tesla's flagship sedan has a driving range of up to 412 miles with a battery pack of 98 kWh. The top-end Tesla S Plaid model has a monstrous combined power output of 1,020 hp with three electric motors!

Tesla Model X: One of the first electric SUVs ever made, it comes standard with two electric motors churning out a total power output of 670 hp. The 3-motor high-performance version called Plaid belches out 1,020 hp. The driving range is 351 miles on the standard and 335 miles on the Plaid variant.

Wuling MINI EV

Volkswagen ID.4

BYD Qin Plus EV

Li Xiang One EREV

Wuling's HongGuang MINI EV: This is China's best-selling electric car, and it costs just between $4,000 to $6,000. It is a 3-door electric microcar featuring an electric motor providing 27 hp and 62.7 lb-ft torque to the rear wheels. It comes either with a 9.2 kWh or a bigger 13.8 kWh with a range of 110 miles.

Volkswagen ID.4: With a single rear-mounted motor, the ID.4 makes 201 hp, but the variant with dual-motor pumps out 295 hp. The battery pack of 77 kWh is capable of a 280 miles range between two charges.

BYD Qin Plus: The full version of Qin (pronounced 'Chin') Plus features a 71.7 kWh battery with a 600 km range. There are also options with 47.5 and 57 kWh battery capacities of 400 and 500 km, respectively. The standard electric motor churns out 136 hp of power.

Li Xiang One EREV: This is a luxury mid-size SUV, a PHEV, or an extended-range EV. It has two electric motors: a 100 kW (130 hp) motor in the front and a 140 kW (190 hp) motor in the rear. It also has a front-mounted 1.2-liter turbocharged 3-cylinder petrol engine. The total power output is 240 kW (320 hp) and 530 N·m (391 lb-ft). The company claims an NEDC range of 700 km (430 mi).

BYD Han EV

BYD Song Plus EV

BMW iX

Ford Mustang Mach-E

BYD Han EV: This mid-size luxury sedan has a power output of 218 hp and a battery capacity of 77 kWh. Depending on the variant, the driving range is between 610 to 715 kms.

BYD Song Plus EV: This is a crossover SUV equipped with a 72 kWh battery pack, giving it a range of 505 kms. It gets its power from a single permanent magnet synchronous motor with an output of 184 hp

BMW is: This is the Bimmer's first battery-electric SUV. Two electric motors provide all-wheel drive and more than ample power

—516 hp in the xDrive50 and a whopping 610 hp in the M60—while a large battery pack of 105.2 kWh delivers enough juice for an estimated driving range of over 300 miles per charge.

Ford Mustang Mach-E: It is available with either a standard-range 70 kWh battery or an extended-range 91 kWh pack. These feed an electric motor mounted on the rear or both axles. That latter combination creates all-wheel drive, and on the sporty GT and GT Performance models, the engines combine to make 480 horsepower and up to 634 pound-feet of torque. The Mustang Mach-E has an EPA-estimated range of between 211 and 305 miles, depending on the battery pack and type of electric motor.

Hyundai Ioniq 5

Hyndai Kona Electric

Chevrolet Bolt EUV

BYD Han EV

Hyundai Ioniq 5: The Ioniq 5 line-up spans a 168-hp, single-motor, rear-drive version with an EPA-estimated 220 miles of range up to a 320-hp, all-wheel-drive, dual-motor unit with an estimated 266 miles of range. The big-range, single-motor, rear-drive model has 303 miles of range. The standard-range battery paired with a single, rear-mounted electric motor generates 168 horsepower. With the longer-range battery, the rear-drive configuration generates 225 horsepower and 258 pound-feet of torque. The all-wheel-drive version is much more powerful, with 320 horses and a substantial 446 pound-feet. The standard battery pack is 58 kWh, while the larger one is 77.4 kWh capacity.

Hyundai Kona Electric: The Kona Electric gets its power from a 201-hp electric motor, and its large battery pack of 64 kWh is good for up to 258 miles per charge.

Chevrolet Bolt: This model gets driven by a 200 hp electric motor and has a 66 kWh battery pack. It can go up to 259 miles on a single charge. As you can observe, the Kona and Bolt have nearly the exact specifications.

Nissan Leaf: The standard Leaf S comes with a 147-horsepower electric motor and a 40.0-kWh battery pack, and the leaf SV Plus gets a gutsier 214-hp electric engine and a larger 62.0-kWh battery. The Leaf S has a relatively limited driving range of 149 miles, while the Leaf SV Plus has a better range of 215 miles.

Kia EV6

Audi E-tron GT

Mercedes-Benz EQS

Polestar 2

Kia EV6: The EV6 is available with a single electric motor mounted on the rear axle that gives it rear-wheel drive or electric motors on both front and rear that enable all-wheel movement. While all-wheel-drive models have a larger battery pack and either 320 or 576 total horsepower, the base rear-drive example has a smaller battery and just 167 horses. Opting for the bigger battery boosts the rear-drive EV6's total system output to 225 horses. The Kia EV6 offers two battery sizes: a 58.0-kWh and a 77.4-kWh pack.

Audi E-Tron GT: It gets its power from a dual motor with a peak power output of 470 hp. The 93.4 kWh battery pack gives it an EPA-estimated range of 238 miles.

Mercedes-Benz EQS: The sizable EV sedan offers a 329-hp rear-drive powertrain (EQS450+) or a 516-hp, all-wheel-drive setup (EQS580 4Matic). Both utilize a large 107.8-kWh battery rated at 350 miles of range.

Polestar 2: As part of Volvo's EV subsidiary, Polestar, this model comes only with a large battery pack and an electric motor powering the front wheels—or two electric motors if you opt for the more powerful all-wheel-drive setup. Unlike the dual-motor, all-wheel-drive variants of the Polestar 2, versions equipped only with the 231-hp single motor are front-wheel drive. The more powerful dual-motor, all-wheel-drive setup features an electric motor on both axles and generates a total system output of 408 horsepower. It accommodates a 75 kWh battery pack with an estimated range of 270 miles.

Volvo XC40 Recharge

Porsche Taycan

Tata Nexon EV

MG ZS EV

Volvo XC40 recharge: With a whopping 402-hp and the instant torque delivery of two electric motors, the Volvo XC40 Recharge is an eco-friendly SUV with some high speed. The model comes with two electric motors, one driving the front wheels and the second powering the rears, giving it full-time all-wheel drive. A 75.0-kWh battery pack is standard equipment, and the EPA has given the XC40 Recharge a range estimate of 223 miles per charge.

Porsche Taycan: From the 402-hp single-rear-motor base model to the absurdly quick 750-hp Turbo S, every Taycan uses a two-speed transmission. The Porsche's ratings range between 199 and 227 miles, depending on the battery size and electric motors.

Tata Nexon EV: The undisputed No.1 electric car (as of now) in India, the Nexon EV has a power output of 127 hp and a rated driving range of 312 km on a full charge. It comes equipped with a 30.2 kWh battery pack.

MG ZS EV: Powered by a permanent magnet synchronous motor, it churns out an output of 174 hp. The battery capacity is 50.3 kWh, giving it a range of 461 km under test conditions.

REFERENCES

Chapter 1: History of Electric Vehicles

- *The History of Electric Vehicles*– www. electricvehicles. com
- *The Fascinating Evolution of the Electric Car* – www.businessinsider.in
- *Electric Car Evolution* by Zachary Shahan – http://www. cleantechnica.com
- *Electric Cars* – www.edisontechcenter.org
- *History of the Electric Vehicle* – en.wikipidea.org
- *History of Electric Cars* – Idaho National Laboratory

Chapter 2: Electric Vehicle Technology

- *Modern Electric, Hybrid Electric, and Fuel Cell Vehicles* – Mehrad Ehsani, Yamin Gao, Sebastien E. Gay, Ali Emadi

- *How Electric Cars Work*– auto.howstuffworks.com
- *The main components of electric vehicles* – yocharge.com
- *EV A to Z Encyclopaedia -1: Understanding EV components* – Hyundai Motor Corp
- *How do all-electric cars work?* – AFDC, US Department of Energy
- *EV 101: How do electric cars work?* – by John Bogna, pcmag.com, June 27, 2022
- *How do electric cars work?* – ackodrive.com, June 9, 2022
- *EV Motors explained* – by Dan Edmunds, caranddriver.com, April 5, 2022
- *Electric Cars 101: How EV motors work, technical differences, and more* – by Frank Markus & Ryan Lugo, motortrend.com, July 8, 2021
- *Everything you need to know about EV motors* – by Stephen Corby, carsguide.com.au, 18 June, 2021
- *Different types of electric motors used in EVs* – by Loukas, arenaev.com, 30 April 2022

Chapter 3: Battery Technology

- *Batteries for Electric Cars – Challenges, Opportunities and the Outlook to 2020* – Andreas Dinger et al., The Boston Consulting Group
- *Battery Electric Vehicles--Performance, CO2 emissions, Lifecycle costs, and Advanced Battery Technology Development*--Daan Bakker, Copernicus Institute University of Utrecht
- *Electric Vehicle Battery Technologies*--Kwo Young,

Caisheng Wang, Le Yi Wang, and Kai Strunz, www.springer.com

- *Electric Vehicles Technology Brief*– www.irena.org, February 2017
- *EV battery technology: The road to a breakthrough* – by Aparna Narayanan, investors.com, May 13, 2022
- *Three battery technologies that could power the future* – by Patrick Bernard, saftbatteries.com
- *An accidental discovery could change the world* – by William Lockett, freethink.com, April 15, 2022
- *Solid-state batteries could be the key for EVs – and they're making a leap towards mass production* – by Dan Gearino, fastcompany.com, June 11, 2022
- *The future of EV batteries* – greencars.com
- *Explainer: Are lithium-ion batteries in EVs a fire hazard?* – by Heekyong Yang, reuters.com, August 24, 2021
- *Here's why EV batteries sometimes burst into flames* – by Loanna Lykiardopoulou, thenextweb.com, January 10, 2022

Chapter 4: Charging Infrastructure

- *Emerging Best Practices for Electric Vehicles Charging Infrastructure*– by Dale Hall & Nic Lutsey, The International Council on Clean Transportation, White Paper, October 2017
- *EV charging speeds* – transportation.gov, US Department of Transportation
- *Trends in charging infrastructure* – Global EV Outlook 2022, International Energy Agency (IEA)

- *Building the electric vehicle charging infrastructure America needs* – Philipp Kampshoff, Adi Kumar, Shannon Peloquin & Shivika Sahdev, mckinsey.com, April 18, 2022
- *Developing infrastructure to charge electric vehicles* – AFDC, US Department of Energy
- *Europe needs 65 million electric vehicle chargers by 2035* – by Ira Boudway, Bloomberg.com, February 8, 2022
- *China to further boost EV charging services* – English.gov.cn, January 22, 2022
- *Vehicle Grid Integration – A Global Overview of Opportunities and Issues* – Nihan Karali & Anand R. Gopal, National Renewable Energy Laboratory, June 2017

Chapter 5: Clean Energy

- *Electric vehicles and renewable energy: a perfect match* – atlasrenewableenergy.com, July 7, 2021
- *The potential of green hydrogen and energy storage* – crswire.com, submitted by Black & Veatch, June 23, 2022
- *What is green hydrogen, and why do we need it? An expert explains* – by Abhinav Chugh & Emanuele Taibi, weforum.org, December 21, 2021
- *India plans to produce five mln tonnes of green hydrogen by 2030* – by Sudarshan Varadhan, reuters.com, February 18, 2022
- *India's green future, built on hydrogen* – by P B Jayakumar, fortuneindia.com, June 9, 2022

Chapter 6: Economics of Electric Vehicles

- *The total cost of ownership and market share for hybrid and electric vehicles in the UK, US, and Japan* – Kate Palmer et al., Applied Energy Journal, Elsevier, November 2017
- *Electric Vehicle Life Cycle Cost Analysis* – Richard Raustad, Electric Vehicle Transportation Center, Florida, February 2017
- *Do electric vehicles save money on gas?* – by Lydia Noyes, leafscore.com, August 29, 2022
- *Here's where it's cheaper to switch to an electric vehicle or not – and how the costs break down* – by Mike Winters, cnbc.com, December 29, 2021
- *Tesla battery replacement cost explained* – findmyelectric.com, March 15, 2022
- *Electric car battery life: everything you need to know* – by Hearst Auto Research, caranddriver.com, April 13, 2020
- *How long should an electric car's battery last?* – by Ceyhan Cagatay, myev.com

Chapter 7: Impact of Electric Vehicles on the Environment

- *The Environmental Benefits of Electric Vehicles as a Function of Renewable Energy* – by Ryan Cornell, Harvard University, May 2017
- *Factcheck – How electric vehicles help to tackle climate change* – by Zeke Hausfather, carbonbrief.org, May 13, 2019

- *Electric Vehicle Myths* – epa.gov, US Environmental Protection Agency, June 30, 2022
- *Are electric cars better for the environment?* – rac.co.uk, March 18, 2021
- *Electric cars and pollution: facts and figures* – virta.global, May 16, 2022
- *Does an electric vehicle emit less than petrol or diesel?* – Lucien Mathieu, transportenvironment.org, April 21, 2020

Chapter 8: Impact on Oil Demand

- *Electric Vehicle Outlook 2022* – Bloomberg New Energy Finance, BNEF
- *BP Energy Outlook* – 2022 edition, BP plc
- *2021 World Oil Outlook 2045, 15th Edition* – Organization of the Petroleum Exporting Countries (OPEC)
- *2*- and 3-wheel EVs are crushing oil demand – by Lisa Martine Jenkins, protocol.com, May 19, 2022

Chapter 9: Effect on Automobile Dealerships

- *A turning point for US auto dealers: the unstoppable electric car* – by Maximilian Fischer et al., mckinsey.com, September 23, 2021
- *How EVs could affect dealers' service department business models* – by John Evans, autocar.co.uk, February 23, 2022
- *The rise of electric vehicles* – by Bob Gray, bdo.com, February 2022
- *Electric cars are here*: what does it mean for dealerships?

– by Dejuan Dickinson, leslieporterfield.com, December 14, 2021

Chapter 10: Tesla Inc

- *History of Tesla Inc.* – Wikipedia
- *The Making of Tesla: Invention, Betrayal, And the Birth of the Roadster*– Drake Baer, www. business insider.in, November 11, 2014
- *The Story behind Tesla's Success* – Amy Wu, Investopedia, July 6, 2016
- *How Tesla Changed the Auto Industry Forever* – Andrew J. Hawkins, www.theverge.com, July 28, 2017
- *Is Tesla a Disruptor? (And why the answer matters)*– Larry Downes & Paul Nunes, Harvard Business Review, August 2017
- *Master Plan, Part Deux* – Elon Musk, Tesla Blog, July 20, 2016

Chapter 11: Legacy Automakers in Electric Vehicles

- *Global Carmakers now target $515 billion for EVs, batteries* – by Paul Lienert & Tina Bellon, reuters.com, November 10, 2021
- *The world's top 5 automakers and their EV plans* – by Junaid Shah, saurenergy.com, February 28, 2022
- *Honda joins electric car race with $40 billion investment* – by Michelle Toh, edition.cnn.com, April 12, 2022
- *Volkswagen increases spending on EVs to $100 billion* –

by Peter Valdes-Dapena, edition.cnn.com, December 9, 2021

- *Toyota has a new EV strategy. Here's where we think it's headed* – by John Voelcker, thedrive.com, January 25, 2022
- *Renault-Nissan to do more together in $26 bln electric bet* – by Tim Kelley & Gilles Guillaume, reuters.com, January 27, 2022
- *The road to 2030* – alliancernm.com, Renault Nissan Mitsubishi
- *Mercedes-Benz strategy update: electric drive* – group.merdes-benz.com, July 22, 2021
- *BMW reveals their new 'factory' strategy* – by Chris Randall, electricdrive.com, May 2, 2022
- *BMW Neue Klasse Platform: specs, features, models & timelines* – by Sagar Parikh, topelectricsuv.com, August 31, 2022
- *Accelerating the drive to electrification* – stellantis.com
- *Hyundai announces accelerated electrification strategy* – by Mark Kane, insideevs.com, March 3, 2022
- *Audi announces 'Vorsprung 2030' electrification strategy* – by Scooter Doll, electric. co, August 26, 2021
- *Hyundai Motor accelerates electrification strategy, targeting 7% of global EV market by 2030* – hyundaimotorgroup.com, March 2, 2022
- *Hyundai, Kia plan $16.5 billion spend to boost EV plan* – by Shinhye Kang, Bloomberg.com, May 18, 2022

Chapter 12: Global Outlook for Electric Vehicles

- *BNEF Electric Vehicle Outlook 2022* – Bloomberg New Energy Finance
- *Global EV Outlook 2022*– International Energy Agency (IEA)
- *Electric Vehicles may control half of the major markets by 2030; Auto Execs predict* – by Keith Naughton, Bloomberg.com, November 30, 2021
- *Global EV Policy Explorer* – International Energy Agency (IEA), May 23, 2022

Chapter 13: China

- *What differentiates China's EV market from the others* – by Parul Chhabra, nickelinstitute.org, April 4, 2022
- *China rises to take half of the global EV market* – by Kim Feng Wong, energyintel.com, January 19, 2022
- *How big is China's electric vehicle market really ?* – focus.cbbc.org, April 6, 2022
- *China: EVs defy Covid to reclaim triple-digit growth* - by Kim Feng Wong, energyintel.com, June 15, 2022

Chapter 14: Norway

- *Perspectives on Norway's Supercharged Electric Vehicle Policy* – Erik Feigenbaum, Environmental Innovations and Societal Transitions, Elsevier, November 2016
- *Why Norway leads the world for electric vehicles* – by Nick Ferris, energymonitor.ai, January 17, 2022

- *What Norway can teach the world about switching to electric vehicles* – by Christina Bu, time.com, January 7, 2022
- *Exploring the electric vehicle revolution in Norway* – by Christina Bu, innovationnewsnetwork.com, May 24, 2022
- *Let's learn from Norway: How Governments can accelerate EV's road to the mainstream* – Tom Willerton-Gartside, ibexpub. media, March 1, 2022

Chapter 15: India

- *Electric Vehicle Industry in India: Why Foreign Investors should pay attention* – by Naina Bhardwaj, India-briefing.com, July 1, 2022
- *Unlocking India's electric mobility potential* – by Barnik Chaitran Maitra et al., adlittle.com, August 2022
- *EV sales in India to breach the 9 mn mark by 2027 – IVCA-EY-IndusLaw report* – by Akashdeep Baruah, businesstoday.in, July 27, 2022

Chapter 16: The other electrics

- *Two-wheelers on a steeper path to zero-emission by 2050* – BNEF Electric Vehicle Outlook 2022, about.bnef.com, June 14, 2022
- *Scooter Story: Asia's Two Wheel EV transition* – by Arvind Sharma, icg.citi.com, December 17, 2021
- *India's 2W industry approaches a crossroad; will electrification takeover* – by Nabeel A Khan, ETAuto, September 9, 2022

- *Bubble alert: India's electric two-wheeler industry may be headed towards a glut by 2026* – by Sumant Banerji, ETAuto, January 20, 2022
- *Electric 3Ws make 47% of all three-wheelers sold in India in FY 2021-22* – evreporter.com, May 0, 2022
- *The electric bus, main fleets, and projects around the world* – sustainable-bus.com, July 12, 2022
- *Global EV Outlook 2022*– International Energy Agency (IEA)

Chapter 17: Electric Car Buyer's Guide

- *Edmunds' EV buying guide* – Edmunds.com, last updated May 5, 2022
- *Electric car buying guide: things to consider before buying an electric car* – acko.com, August 4, 2022
- *9 essential tips for buying an electric car* – by Tom Pritchard, tomsguide.com, August 22, 2021
- *The beginner's guide to buying an electric vehicle* – by Paul Hudson & Alex Robbins, telegraph.co.uk, August 10, 2022

Chapter 18: World's bestselling electric cars

- *Ten most popular electric cars* – by Ben Stewart, kbb.com, July 27, 2022
- *Best electric vehicles for 2022, tested* – caranddriver.com, June 24, 2022

Websites referred

- Ev-volumes.com
- Insideevs.com
- Cleantechnica.com
- Iea.org
- About.bnef.com

Printed in Great Britain
by Amazon

35665447R00169